T0108102

EVOLUTION

WHAT EVERYONE NEEDS TO KNOW®

EVOLUTION

WHAT EVERYONE NEEDS TO KNOW®

ROBIN DUNBAR

OXFORD
UNIVERSITY PRESS

OXFORD
UNIVERSITY PRESS

Oxford University Press is a department of the University of Oxford. It furthers the University's objective of excellence in research, scholarship, and education by publishing worldwide. Oxford is a registered trade mark of Oxford University Press in the UK and certain other countries.

"What Everyone Needs to Know" is a registered trademark of Oxford University Press

Published in the United States of America by Oxford University Press 198 Madison Avenue, New York, NY 10016, United States of America.

© Oxford University Press 2020

All rights reserved. No part of this publication may be reproduced, stored in a retrieval system, or transmitted, in any form or by any means, without the prior permission in writing of Oxford University Press, or as expressly permitted by law, by license, or under terms agreed with the appropriate reproduction rights organization. Inquiries concerning reproduction outside the scope of the above should be sent to the Rights Department, Oxford University Press, at the address above.

You must not circulate this work in any other form and you must impose this same condition on any acquirer.

Library of Congress Cataloging-in-Publication Data
Names: Dunbar, R. I. M. (Robin Ian MacDonald), 1947– author.
Title: Evolution : what everyone needs to know / Robin Dunbar.
Description: New York, NY : Oxford University Press, [2020] |
Series: What everyone needs to know |
Includes bibliographical references and index.
Identifiers: LCCN 2019042379 (print) | LCCN 2019042380 (ebook) |
ISBN 9780190922894 (hardback) | ISBN 9780190922887 (paperback) |
ISBN 9780190922917 (epub)
Subjects: LCSH: Evolution (Biology) | Evolution.
Classification: LCC QH366.2.D857 2020 (print) |
LCC QH366.2 (ebook) | DDC 576.8—dc23
LC record available at https://lccn.loc.gov/2019042379
LC ebook record available at https://lccn.loc.gov/2019042380

1 3 5 7 9 8 6 4 2

Paperback printed by LSC Communications, United States of America
Hardback printed by Bridgeport National Bindery, Inc., United States of America

CONTENTS

3 Evolution and Genetics   59

4 Evolution of Life   82

5 Evolution of Species 104

6 Evolution of Complexity 129

EVOLUTION

WHAT EVERYONE NEEDS TO KNOW®

PREFACE

In the last paragraph of his book *On the Origin of Species*, Charles Darwin referred to life on earth as "an entangled bank"—a collection of different species whose lives are complexly intertwined, sometimes competing, sometimes interdependent, sometimes cooperating. Of course, not that long ago, the tangled bank of nature was a great deal richer in its plants and wildlife than it is now—thanks mainly to what we have done to it in a frighteningly short span of time. But that's another story. For now, the more important point is that our planet is still amazingly rich, teeming with life forms that exist in some remarkably complex relationships, both with one another and with the physical planet itself. It might even be the richest planet in the universe, or at least so we often like to think. But even if there are other planets elsewhere that are just as rich, or even richer, the very fact that it is as rich as it is should excite our curiosity and demand an explanation. How on earth did it come to be like this?

After two centuries of intensive scientific effort, we now have the luxury of a theory that provides a general explanation for that richness, often in quite considerable detail. That theory, Darwin's theory of evolution by natural selection, is famous for two reasons. One is that it is the second most successful theory in the history of science (after quantum theory in physics) in terms of its ability both to explain what we see in the natural world and to stimulate new ideas and research that have uncovered rich seams of novel findings. The second has been its ability, as a theory, to provide a unifying framework for a disparate array of disciplines that do not always see themselves as natural allies. That array includes not just the various life sciences (ecology, genetics, anatomy, physiology, biochemistry, and animal behavior), but also "hard" sciences like chemistry, the softer sciences like medicine, sociology, anthropology, and economics, and even the humanities. History, linguistics, and literature all fall under the purview of evolutionary theory.

At the same time, the theory of evolution is perhaps the most widely misunderstood theory in the whole of science— ironically, it is sometimes misunderstood even by other scientists as much as by laypeople. Of course, the theory we have now is not just Darwin's original theory. That theory has been much developed and extended, to the point where Darwin himself would not recognize large parts of the modern theory as his own—though he might well be intrigued and impressed by it. Nonetheless, everything in the modern theory derives from Darwin's original insights.

My own interest in evolution stems from an unusually early introduction. When I was aged about eleven and living in Africa, my grandmother in California (a retired surgeon with a deep interest in the natural world, but equally deep religious views) used to send me the Audubon Society's "sticker" books for children. I confess that my eyes tended to glaze over with the ones on the desert plants of the American Southwest or the natural wonders of the world. But I found the one on evolution especially captivating, not least because it dealt with the whole of our planet's history, with sections on the dinosaurs as well as human evolution. I have no doubt that this small evolutionary moment of grandparental solicitude probably did more than anything else to encourage me to become interested many years later in both human evolution and wildlife biology, and, in due course, to be willing to spend many hours watching animals, as well as humans, in the wild. The combination of the pleasure that derives from watching other species at their everyday work and the curiosity to ask why and how they come to be as they are and do what they do are both rewards in themselves and the principal drivers of curious inquiry.

Although the theory of evolution itself is elegantly simple—something that all theories in science aspire to be—applying it in a multidimensional substrate like the biological world results in the generation of extraordinary complexity rather as the ripples on a pond spread beyond the point where the stone lands. This is a consequence of the fact that evolutionary processes are embedded in the physics and chemistry of the planet and the ways these interact with physiological and cellular processes, as well as with the cognitive and behavioral machinery that is made possible by the evolution of large brains. I like to think that evolutionary biologists sit in front of an enormous jigsaw puzzle, whose myriad bits at first seem chaotic and completely unrelated to each other. But, as the picture is gradually assembled, something emerges that is both coherent and magical in the way the pieces slot into place. The best science is always about that "I would never have thought that" response as the mists of our confusion clear and all is revealed.

This book does not particularly seek to justify the theory of evolution, or prove that it is true. I take that largely as read. In any case, there are plenty of books that do that already, not least among them Jerry Coyne's recent volume *Why Evolution Is True*. Instead, I want to take a much broader view than most books on evolution would take. I want to show that if we keep asking "But *why* is this the case?," Darwin's ideas turn out to have implications for every aspect of life on earth, including every aspect of human behavior. In doing so, I build on many decades of research in a very wide range of disciplines. I will try to show how the jigsaw of disciplines articulates in ways that support and reinforce Darwin's original insights, creating a satisfying sense of completeness.

Perhaps more than anything else, though, my concern will be with the importance of asking the right questions. Too often, misunderstandings about evolutionary theory and its implications for human behavior arise out of a failure to understand exactly what it is that evolutionary theory actually entails. Nothing could illustrate this better than the bizarre sociobiology wars of the 1970s when the first faltering attempts to apply evolutionary theory in a serious way to human behavior provoked a ferociously defensive backlash from social scientists and the humanities, most of whom seldom took the trouble to find out exactly what was being suggested—let alone read a book on evolution. We have moved on, but I continue to be surprised by the frequency with which the same errors and misunderstandings reappear in different guises—and reflect a continuing failure to engage with the theory.

This book, then, tries to set out exactly what the theory of evolution entails. Most books on evolution, including those childhood Audubon books, deal only with conventional topics like species evolution, the questions Darwin himself had started with—namely, how do new species arise, and why do some go extinct? I want to follow Darwin in his later—in many ways more interesting—explorations of the evolution of behavior, emotions (psychology), and sociology. As Darwin recognized, his theory of evolution has far-reaching implications for every aspect of life, and almost every other discipline in the academy. That inevitably takes us beyond what animals are, to what they do, and so eventually to the human condition itself in all its many, varied aspects.

What the theory of evolution offers us is the greatest story ever told: how we and all the other creatures with whom we share the planet came to be, why we are not all the same, and how and why we are all so interdependent. One of the important messages that an evolutionary approach gives us, and that I really want to emphasize, is that the biological world is a complex, integrated system, not a set of unrelated causes and their unitary effects. Everything an organism chooses to do or become has consequences that reverberate throughout its biology, and even beyond into the biology of other species. Everything we, or any other animal or plant, choose to do has consequences for the evolution of every other member of our species, and for every other species, even every ecosystem, that our actions impact on, as conservationists have been reminding us with increasing alarm.

In some cases, a consequence may throw up constraints that are so severe or so costly to overcome that evolution along these lines is simply impossible. There is, as the behavioral ecologist Nick Davies once remarked, no way butterflies could ever evolve machine guns, no matter how advantageous these would be in their sometimes spectacular aerial battles for territories. Flight and a miniature body size simply rule out anything like that. But even practical innovations will have consequences that the animal will need to address. Growing a larger brain so as to effect smarter solutions to the business of everyday survival means that you will need to forage more, because brains are 10 times more costly to maintain than muscle. That in turn will expose you to higher levels of predation risk. A solution to a problem very quickly becomes another problem that also has to be solved. In evolutionary analyses, we forget this at our peril.

Evolution is also about serendipity: casual circumstance often triggers a direction that could never have been anticipated—the law of unintended evolutionary consequences. One example close to home was, as we shall see later, the adoption of bipedalism by our earliest hominin ancestors six or seven million years ago: this made it possible for speech to evolve some six million years later. Neither our ancestors nor evolution intended this consequence, but it was an opportunity that was, at the right moment, there for the taking (though it did, as we shall see, require some other serendipitous and quite unrelated adaptations along the way). Had our distant ancestors not been pushed down this road, it is very unlikely that human language (and all that followed from that) would ever have evolved. I would not have been able to write this book, and you would not be reading it.

In telling this story, I have structured the book as a conversation that builds progressively, as a conversation might. The key to it, as to all science, is the child's persistent asking in response to every answer, "But why?" The 10 chapters identify 10 major topic areas that progressively narrow the focus down from evolution in the abstract to human cultural behavior in particular. I want to show how by simply asking "Why?" we are taken inexorably down into every corner of life. I have necessarily been somewhat selective in my topic choices, but then so has every book on evolution. My choices have been dictated mainly by the way key parts of the story build progressively into a coherent explanation for much of what we, as humans, are, do, and experience— humans as just another species, but one that, for better or worse, we happen to have a particular interest in. Above all, though, I want to convey the sense that science, difficult as it sometimes can be, is enormous fun, not least because it can often surprise us in a way that our folk beliefs seldom do— because they rarely invite us to be curious about the world "out there." It is a story that never ceases to amaze and intrigue, and that, like all the best stories, never seems to pall in the retelling. Indeed, in many ways, science is one of the pinnacles of human storytelling, and the theory of evolution is one of its best examples.

I should add, just to be clear, that I will often speak of animals intending to pursue particular evolutionary strategies, or of genes promoting their own self-interest. This does not imply that either genes or animals (or, indeed, humans sometimes) are conscious of what they are doing, or even that they are intending to act in a particular way. This is simply a convention in evolutionary biology, which derives from the fact that natural selection acts *as if* (in a metaphorical sense) it was selecting for animals to behave in a teleological manner. One could, if so minded, rephrase everything in a much more precise way, but it would inevitably be long-winded and make for a very dull read. In science, nothing is gained by making things more complicated than necessary, not least because the human mind is not designed to handle complexity.

1

EVOLUTION AND
NATURAL SELECTION

1. Why do we need a theory of evolution?

We live in a world of immense diversity. There are around 5,500 species of mammals and some 10,000 species of birds. But even that pales into insignificance beside the 12,000 species of roundworms, about half of whom are parasites—the ascarids, hookworms, pinworms, and whipworms that play havoc with human health all around the world. And these in turn are dwarfed by the 91,000 species of insects that we currently know about—never mind an estimated similar number that we haven't yet identified. And we haven't even considered the 250,000 species of beetles. As the great English evolutionary biologist J. B. S. Haldane once observed: if God exists, he must be inordinately fond of beetles. On top of that, you can add around 250,000 species of plants. And we haven't got on to the bacteria and viruses yet. . . .

Most of the time, we take all this extraordinary diversity for granted, hardly noticing it as we go about our daily lives. Typically, we are only interested in species that cause diseases, are a nuisance in our gardens, or appear on our dinner plates. The numbers that fall into these categories are so small as to be insignificant: most of us eat no more than 100 species of animals and plants combined, and many a good deal fewer than this.

Given that we share our world with so many species, it should give us pause to wonder how they—and we—came to be. It is easy to look at the world as it is now and imagine that this is how it has always been. Even the great Greek philosopher-scientist Aristotle (384–322 BC) assumed the world had always been as he found it—despite being, by a clear head and shoulders, by far the greatest biologist before Darwin.

Even so, we are left with a puzzle: why are there so many *different* forms of life? Why should there be so many species of beetles, so many birds, so many different plants? On the Noah's Ark principle, wouldn't one species of beetle be enough? The very fact that there are hundreds of thousands of different forms of life raises another important question: why are some of these life forms beneficial for us (for example, the ones we eat or those that coexist happily with us) whereas others (such as the many species of predators and parasites) are decidedly detrimental to our health and even our very survival? The answer to all these questions lies in Darwin's theory of evolution by natural selection.

The great Ukrainian-American evolutionary biologist Theodosius Dobzhansky (himself a devout Orthodox Christian) famously remarked, "Nothing in biology makes sense except in the light of evolution." He might have said "Nothing in *life* . . .," because Dobzhansky's dictum, as this is now known, applies as much to the human sciences—psychology, sociology, economics, history, archaeology, and almost any other discipline you care to name—as it does to the biological world. Everything in our world has evolved, just as the solar system and the universe itself have evolved. Nothing, not even the universe, has remained in a stable state since the beginning of time. Everything has a history dominated by change with the progression of time. That is all that evolution means.

But this raises one more question: is that change simply the natural unfolding of an internal process of some kind, or is it the consequence of some kind of external force? No one doubts that the evolution of the universe is of the first kind.

Early theories of biological evolution dating back to the ancient Greek philosopher-scientists also assumed the first, but Darwin insisted that it was the second. The question we have to ask is: how is it that we have ended up with the biological and psychological world as we now experience it? And how does this impinge on us, as a species, and on our behavior?

2. So who discovered evolution?

The great classical and medieval philosopher-scientists accepted that the various life forms differed in terms of their skills and abilities—a view first set out by Aristotle around 350 BC. Building on Aristotle, the Christian theologians of the early medieval period developed an implicit theory of evolution known as the Great Chain of Being. At the top of the chain was God, with his angels on the rung immediately below Him, and then, on the rung below that, humans—who, everyone agreed, are the most advanced form of life on earth. And then on down, rung by rung, through the various mammals, birds, reptiles, and fishes to the bugs and creepy crawlies that represent the least developed forms of life.

It was easy to view this as a static hierarchy, fixed in time by an act of creation. However, during the 1600s, people began to wonder about some of the rocks that farmers occasionally dug up when plowing their fields. They were clearly rocks, but advances in medical science, especially anatomy, made it obvious that these rocks looked suspiciously like bones. In 1667, the Danish priest and scientist Nicolas Steno (or Nils Steensen in the original Danish) argued that they must be the bones of long-dead creatures that had become rocklike through the absorption of minerals while in the ground.

This surprising suggestion raised the possibility that species might not live forever—that some species might go extinct. That realization sowed the seeds of the idea that life forms might evolve and change over time. Trying to explain how this happens, however, proved to be a challenge. Many zoologists during the 1700s and early 1800s made suggestions. The most successful of these theories was that proposed in 1809 by the most eminent of all the French zoologists, Jean-Baptiste de Monet, the chevalier de Lamarck[1]—or plain Lamarck as he is now usually known.

Lamarck supposed that species were constantly being created anew—even now, this very minute, right now. Each of these new species then evolved along the same set path—the Great Chain of Being. Everyone begins as a tiny bug of some kind and gradually works their way up the Great Chain until eventually everyone presumably becomes human. (It was never quite clear whether we as a species would one day become angels, or whether it meant that new human species could evolve while we were still around.) One important implication was that the species that have been around longest are the most advanced because they have had the most time to work their way up the Great Chain. This implied that humans, being obviously the most complex species, must have been around the longest—notwithstanding the clearly stated fact in the Bible that Adam and Eve had in fact been created *after* all the other beasts of the field. Awkward, but something that everyone seemingly chose to ignore.

Though these views owed as much to armchair philosophizing as real empirical zoology, the Lamarckians in fact had good experimental evidence to support their view. The experiment was deceptively simple. Take a bucket of water from a pump, check it carefully under a microscope, and you'll see nothing but pure water. Leave the bucket to stand for a few weeks and check it again—and, lo and behold, you will find it teeming with microscopic life. Here was startling evidence that life was constantly—and spontaneously—being created anew every minute of the day.

It was an ingenious experiment. Its only flaw was that knowledge about germs and bacteria was still a century away in the future. The "newly created" life was, of course, coming in from the surrounding air, which, as we now know, is full of microscopic life blown hither and thither on the wind.

Meanwhile, in Britain, a number of amateur naturalists had also been promoting evolutionary views. These included Erasmus Darwin (Charles Darwin's grandfather), the Edinburgh publisher Robert Chambers, and the Scottish farmer and founding father of geology James Hutton (who proposed something that was remarkably close to Darwin's theory of evolution by natural selection). Darwin likely came across these views when he was a medical student at Edinburgh (before he gave that up in favor of becoming a theology student at Cambridge with a view, at his father's exasperated behest, to becoming an Anglican parson).

3. How did Darwin's theory change our understanding of evolution?

In 1831, a very young Charles Darwin (he was just 22 at the time) was offered the opportunity to accompany an equally young captain, Robert FitzRoy (he was just 26),[2] on a five-year, round-the-world voyage on the Royal Navy's survey vessel *HMS Beagle* to survey the coasts of South America and the Pacific islands as far as Australia. The purpose of this lengthy voyage was to improve the quality of the Royal Navy's shoreline maps. Darwin's role was very simple: it was simply to be FitzRoy's companion on the voyage so that the young captain had someone of his own intellectual standing to talk to on what would otherwise be a very long, lonely trip. But, being already an experienced geologist, Darwin was soon appointed to be the expedition's geologist. In fact, he spent much of his time undertaking long horseback rides into the hinterland at every stop as an escape from the boredom of shipboard life. During these expeditions, he collected the fossil bones of extinct giant South American mammals as well as specimens of the many new living species of insects, birds, and mammals that he encountered.

The similarities and differences between the species he collected in different places caused Darwin to spend the next 25 years grappling with how some of these species, and especially those on neighboring islands, could be so different yet so obviously related. Eventually, in November 1859, he published his groundbreaking book *On the Origin of Species by Means of Natural Selection* in which he set out the case for a new theory of evolution.

Darwin's theory was a complete contrast to the Great Chain of Being theories. So far from assuming that life is being created anew all the time as the Lamarckians did, he proposed a single ancient origin to all living and extinct species, all of whom were thus related to each other by descent from a series of common ancestors. Complexity and being "advanced" had nothing to tell us about how new or old a species was; these simply represented the history of challenges that a species had encountered from the various environments to which it had been exposed. More importantly, species didn't all follow the same evolutionary trajectory, as the Great Chain theories assumed; instead, each followed its own individual trajectory through time that depended on the habitats it happened to colonize, the opportunities that came its way, and chance events like volcanic eruptions and hurricanes.

What made Darwin's theory more successful was that it provided a mechanism that could drive evolution. This was his concept of natural selection and the idea of adaptation that it naturally gave rise to. In fact, Lamarck's theory had had its own concept of adaptation, now known as the *inheritance of acquired characteristics*, which argued that the massive arm muscles that blacksmiths developed from constantly hammering horseshoes would in due course be passed on to their children. How this was possible was never quite clear, but somehow the trait must become part of their biological makeup. Darwin argued that changes in a species' appearance arise only because individuals vary naturally in the extent to which they exhibit a trait, and only those traits that are best adapted to the local circumstances reproduce successfully; as a result, over many generations a species will come to resemble one particular ancestor more because its particular traits are better adapted to the local environment. We might sum up the differences between the two theories in this way: Lamarck imagined that use gives rise to variation, whereas Darwin imagined that variation gives rise to use.

Darwin was not the only person thinking along these lines at the time. In many ways, the Scottish grain merchant Patrick Matthews had worked out a similar theory of natural selection while trying to breed more productive crops. Indeed, Darwin's views had been much influenced by the success of the animal and crop breeders of the agricultural revolution of the late eighteenth century. More importantly, the explorer-naturalist Alfred Russel Wallace had quite independently come up with a very similar theory while collecting animals and plants in South America and the Far East. Famously, it was Wallace's letter to Darwin in 1857 enclosing a summary of his ideas on natural selection that forced Darwin's hand into publishing his own theory.

Although Wallace deserves great credit for independently arriving at the same explanation as Darwin—and doing so, not in the comfort of a large country house in Kent, but under the very trying conditions of tented camps in Borneo—it is Darwin who gets the full credit because of the rich detail of the evidence that he provided along with a more sophisticated understanding of the mechanisms of inheritance. Moreover, Wallace was never quite willing to abandon the idea that God was involved somewhere—and, in fact, spent much of his later life back in England trying to justify a role for God in evolution. Darwin's theory had no need for God, and so was simpler and more elegant as a scientific theory.

In many ways, Darwin's achievement was to ask the big question, marshal evidence from a wide range of scientific disciplines such as geology, anatomy, and ecology, combine this with his own observations from all around the world, fit all these bits of the jigsaw together so he could recognize a pattern, and, having done all that, finally put together an explanation that no one had thought of before in a way that even a layperson could understand. By the same token, of course, although we can speak of Darwin's theory as the particular set of ideas that he articulated in the second half of the nineteenth century, biologists have continued to develop and extend his theory in the century and a half since his death. This modern theory exhibits the marks of Darwin's original ideas, but it is hugely more complex and sophisticated—to the point that there are many aspects of the modern theory that would probably surprise and amaze Darwin. That said, however, it is important to appreciate that most of these new ideas, concepts, and facts would probably never been discovered if biologists had not had Darwin's theory as a framework for asking questions and testing hypotheses.

4. What was Darwin's theory of evolution by natural selection?

Darwin's theory can be summed up in the form of three simple propositions, or axioms (or assumptions, if you prefer), and the logical conclusion that follows from them:

   i. Individuals vary in the extent to which they possess a given trait [the principle of variation];
  ii. some of that variability is genetically inherited [the principle of inheritance];
 iii. some variants are able to reproduce more successfully (because they are better suited in some way to the environment) [the principle of adaptation];
and hence, as a direct consequence,
  iv. the next generation will come to resemble the more successful variants [the principle of evolution].

Darwin's ideas were founded on the observation, enunciated by the Rev. Thomas Malthus in a very influential book published in 1798, that the rate at which a population can give birth will eventually always outstrip the habitat's capacity to feed it, leading inevitably to the premature death of many individuals.[3] Darwin read Malthus's book while on the *Beagle*, and it later provided him with the spark for the idea of natural selection—that because not every individual will survive, Nature will act like a sieve, allowing individuals who are better adapted to local conditions to survive and reproduce more successfully than those who are less well adapted.

Darwin's theory was, in effect, an amalgam of three separate processes (inheritance, selection, and evolution), each of which has to be true for evolution to happen. He viewed the selection process as being the most important. Natural selection is not the same thing as evolution. Evolution is what happens when natural selection acts on organisms that have heritable traits. More importantly, the relationship between these two components is not symmetrical. If natural selection occurs, evolution is inevitable, but evolution can occur in the absence of natural selection.

Darwin's point was that natural selection provides the sieve that greatly speeds up the rate *and* direction of evolutionary change within a lineage. In the absence of natural selection, evolution would be slow, random, and undirected. Two populations will gradually drift apart as a result of the chance accumulation of minor differences, but evolution will be a very slow process (see Question 29). Natural selection can greatly speed this process up and push a species toward a particular end point. However, unlike the Great Chain of Being, there is no inevitable end point where every species will end up. In Darwin's conception of evolution, the end points are unpredictable: they always depend on accidents of circumstance and the serendipity of opportunity along the way. No two evolutionary pathways will ever be exactly the same.

It is important to remember what Darwin was trying to explain: why life on earth is so diverse, why there are so many different species, and why some of these species seem to be very similar to each other yet others so different. In particular, he wanted a mechanism that could explain *rapid* evolutionary change—rapid, that is, on the geological time-scale. Remember, all this was in a context in which many of Darwin's own academic mentors had been instrumental in demonstrating that the age of the earth was many orders of magnitude greater than people had previously imagined. James Ussher, the seventeenth-century archbishop of Armagh in Ireland, had added up the ages of all the people listed in the Bible's Old Testament books to arrive at a total time elapsed since the day when Adam and Eve were created, and concluded that this had happened in 4004 BC. The geologists had concluded that in fact the earth was several million years older than this. Darwin himself had made seminal contributions to this particular geological story and would still be remembered today as one of the founding fathers of modern geology even if he had never come up with his theory of evolution.

5. What evidence did Darwin have to support his theory?

Darwin drew on three main sources of evidence in developing his theory of evolution by natural selection. One was comparative data on similarities in the anatomy of living species. During the 1730s, the great Swedish taxonomist Carl Linnaeus had shown that species can be clustered together in hierarchical groupings (species within genera, with these then combined into families) based on similarities in their anatomy. Linnaeus made no assumptions about evolutionary history (he was simply a classifier), but it became obvious to the later evolutionists that species that resembled each other were likely to have had a common ancestor (see Question 41). Darwin, like most biologists of his day, accepted this principle and used it to draw attention to the fact that species that lived near each other (such as the finches of the Galapagos Islands and the nearby South American mainland) often resembled each other, whereas those from different continents were often very different.

A second source of evidence was that for local adaptations. Darwin had not been especially interested in the birds that he collected on the Galapagos Islands, but the ornithologists to whom he gave the specimens when he got back to England were struck by the fact that the finches varied greatly between the various islands. Although obviously all were finches, some were small and had very fine beaks, others were large with heavy beaks, with every variation in between. Once Darwin realized this, he was able to work out from his field notes that these differences were adaptations to the different feeding opportunities on the various islands. Adapting to these had resulted in a gradual shift toward different types of beak as the individuals who possessed beaks that were better suited to a particular diet reproduced more successfully.

The third source of evidence that Darwin relied on to support his theory was the evidence from breeding experiments. Having recognized that a species might change its anatomy in response to changes in its environment, Darwin was faced with the difficulty of explaining how this could come about if species were supposedly immutable, as many people assumed. His interest in breeding experiments had been piqued by the agricultural experiments of the late eighteenth and early nineteenth centuries, when breeders like Robert Bakewell and Thomas Coke had managed to develop improved breeds of domestic stock and as well as crops. In addition, pigeon fancying was then a popular pastime and involved the careful pairing of birds to produce new varieties, some of which had dramatically different appearances. Darwin spent a long time studying their methods and trying them out for himself. Not only did he devote a chapter in *The Origin of Species* to the topic, but he later wrote an entire book on the subject (*The Variation of Animals and Plants Under Domestication*).

Eventually, Darwin became convinced that this artificial form of breeding provided evidence for how nature might work to change a species' appearance over time, with natural selection filling the role of the human breeder. Small but consistent changes in the environment might act as the pigeon breeder did, selecting for particular variants within the natural range of variation available in each successive generation. Because not every variant survives to reproduce, the population will gradually come to resemble the most successful parents, just as the pigeon breeder was able to produce birds that increasingly match his vision of the perfect individual.

6. In that case, why did Darwin propose a second theory, the theory of sexual selection?

Darwin was very anxious about weaknesses in his theory, which is one reason he waited so long before publishing it. He wanted to make sure that he had as much evidence to support his claims as was humanly possible. This is also why he devoted so much of his time to studying seemingly irrelevant phenomena like pigeon breeding (see Question 5), pollination in plants (his two books *On the Various Contrivances by Which British and Foreign Orchids Are Fertilised by Insects* and *Insectivorous Plants*), the behavior of earthworms (*The Formation of Vegetable Mould Through the Action of Worms*), the biology of barnacles (*A Monograph of the Cirrepedia*), and the nature of coral reefs (*The Structure and Distribution of Coral Reefs*). Still, one topic eluded him: why did some species have elaborate appendages or coloration that seemingly handicapped them by making it difficult to fly or, in the case of bright colors or loud mating calls, risked drawing the attention of predators?

Darwin eventually realized that natural selection involves two completely separate components (survival and reproduction). It is the way these interact to determine the number of descendants that an individual leaves (the motor of evolution itself) that is important. An organism has both to survive *and* to reproduce, but up to a point it can trade one off against the other. In other words, an animal can afford to die young providing it produces many offspring while it can; if it wants to take reproduction at a more leisurely pace, then it had better take care to survive longer.

The importance of reproduction in this process opens up the possibility that organisms may be able to enhance their own reproductive success by attracting mates that are reproductively more successful. This may be especially important if mating opportunities are rare, and this leads to competition for access to mates. Darwin referred to this as *sexual selection* to differentiate it from natural selection (where "natural" is often interpreted as referring to adaptations for survival). Strictly speaking, of course, sexual selection is just one component of natural selection; the other is what we might call "survival selection." Sometimes, however, sexual selection can result in the evolution of traits that are disadvantageous in terms of mere survival. Being able to produce more offspring than a rival gives one an evolutionary edge. The species will eventually come to resemble the individual that has the most offspring, especially if the reproductive advantage is repeated in generation after generation because it reflects some heritable trait.

Darwin distinguished two different ways in which sexual selection might act: one was through competition between members of the same sex in order to monopolize matings with the opposite sex (intrasexual [or within-sex] selection); the other occurs when members of one sex display to those of the other sex who actively choose whom to mate with on the basis of the traits displayed (intersexual selection [or between-sex] selection).

The first (intrasexual selection) is illustrated by male deer or elephants who fight each other, often in dramatic battles, in order to control access to breeding females. The result is often selection for ever larger body size or ever bigger weapons (such as antlers or tusks). The females, in this case, are simply passive observers, mating with the winner, whoever he happens to be. This is not to say that the females are disinterested participants, however; in effect, they are relying on the fighting to sort out the best males for them. After all, the traits that make a male an effective combatant will presumably be beneficial in due course for her male offspring.

In contrast, intersexual selection is more subtle in that one sex actively seeks to attract the other sex in some way. One familiar example would be the peacock, whose males have evolved large, colorful, and *very* ungainly tails that they display to females in the hopes of impressing them sufficiently to persuade them to mate. In some species, several males display on small territories close to each other (known as *leks*); the females wander around the lek and check out each displaying male before deciding which one to mate with. American prairie grouse (or prairie chicken), with their booming vocal displays and tail-snaps, are another example. In these cases, the traits that the males display are usually correlated in some way with their ability to survive or reproduce. A peacock with an especially large tail in effect says: see how good my genes are—I can escape from predators even when saddled with this enormous handicap, so choose me! For this reason, this process is often known as the *handicap principle,* elaborated in the 1970s by the Israeli biologist Amotz Zahavi. In other cases, females may select for traits in males that are directly related to some aspect of reproduction such as good parenting skills. In some fish, like the freshwater stickleback, the male prepares a nest that he displays to passing females; if one likes the look of it, she will pass through and lay her eggs—after which she heads for the fish equivalent of the pub and leaves the male to do all the work required to hatch the eggs and guard the hatchlings until they are old enough to fend for themselves.

In some cases, however, animals can select for traits that are completely random. This is sometimes known as *Fisher's sexy sons hypothesis* after the founding father of mathematical genetics, Ronald Fisher, who first proposed it in the 1930s: a female will prefer to mate with males that have traits that will allow the female's sons to be more attractive to females. It is also known as *Fisherian runaway selection* because it can lead to the very rapid evolution of completely random traits that females just happen to like, such as the bright blue testicles of some African monkeys or the brilliant curly tails and spectacular displays of New Guinea birds of paradise, or the tuneful singing of many songbirds. In many cases, it does this by picking up on and exaggerating features that are already present for some completely different, but biologically legitimate, purpose. These are said to provide a window of evolutionary opportunity on which sexual selection acts.

Sexual selection has played a particularly important role in species evolution because it can speed up the rate at which two populations separate and become two distinct species (see Question 43). Within species, it can also provide an explanation for unusual or exaggerated aspects of an animal's appearance. In our species, this seems to include men's beards and deeper voices, women's hourglass body shape, and young men's competitive, risk-taking behavior.

7. What evidence do we now have to support Darwin's theory?

There are six sources of evidence for evolution available to us now that weren't available to Darwin in the second half of the nineteenth century: a much richer fossil record, evidence from embryology (how organisms develop), evidence from common biochemical processes, molecular genetics, laboratory experiments on natural selection, and evidence from field studies of animal and plant populations for natural selection at work in nature.

A century and a half of intensive fossil hunting has yielded fossils for a vast array of long-extinct species, as well as fossils for many living species dating across a wide range of time. One consequence of this, for example, has been the recognition that there have been five major extinction events over the past 500 million years, when 70% or more of all species were wiped out by some major environmental catastrophe (see

Question 46). The richer fossil record has also allowed us to piece together the historical sequences by which different species evolved. One of the earliest to be reconstructed was that for the horse family, showing the changes over the last 40 million years from the dog-sized, four-toed *Eohippus* to today's large-bodied horses with their single toe. Similar sequences have been built up for many major groups of animals and plants. More recently, it has been possible to trace the lineages of marine organisms such as long-extinct trilobites and even tiny planktonic radiolarians as they evolved slowly over many millions of years.

The second important source of evidence comes from studying the processes whereby organisms develop (embryology) and comparing these between different groups of animals. In many cases, this has revealed that a wide range of distantly related species undergo the same developmental processes in the same order. In animals with backbones, for example, the neural cord is laid down early on at the same stage in all species. Many species briefly exhibit vestigial organs found in other unrelated families (see Question 15). Indeed, except to an expert, human and dog embryos are virtually indistinguishable until they are about a third of the way through pregnancy.

A third source of evidence is the fact that the biochemical processes on which life depends are identical across a wide range of species from many different families. Some 99% of all living cells, from microbes to humans, depend on just six chemical elements (carbon, hydrogen, nitrogen, oxygen, calcium, and phosphorus). With very minor variants, everything from yeast to plants and humans has the same cytochrome-c (a hemeprotein essential for the transportation of electrons within cells). Indeed, with the exception of one group of archaic bacteria, everything from viruses to humans uses the same set of nucleic acids as the basis for their genetic codes. The likelihood of all of these being the case by chance is infinitesimally small, making inheritance from a common ancestor and a common origin deep in evolutionary time by far the most likely explanation.

Perhaps the most important form of evidence supporting a general theory of evolution comes from molecular genetics. Molecular genetics was born in the late 1940s, and really came into its own in 1953 with the discovery of the genetic code by Francis Crick and James Watson in Cambridge and their colleagues and rivals Maurice Watkins and Rosalind Franklin in London.[4] The strands of DNA that make up our chromosomes (the only thing that passes on from parents to their offspring) consist of sequences of the same four nucleic acid bases arranged in different orders, blocks of which (known as codons, or, more colloquially, genes) are responsible for creating our bodies (see Question 24). From the 1980s onward, it has become possible to compare the complete genetic codes of different species and so build up a taxonomy based on identity of individual DNA bases. In many cases, this confirmed the relationships that Linnaeus and the later taxonomists had inferred from anatomy—but not always! In some cases, there were unexpected surprises (though mostly only at the level of detail) (see Questions 44 and 61).

Knowledge of genetics makes it possible to determine exactly how closely related any two species are, and how long ago they shared a common ancestor. We humans share around 95% of our DNA with chimpanzees, 85% with dogs, and as much as 40% with the humble cabbage. This is a reflection of the fact that most of the basic building blocks of our bodies (bones, blood, nerves) and many of the biochemical processes involved (oxygenation of cells, processing of fats and carbohydrates) are very similar across species.

Finally, experimental studies of selection in the lab and observational studies in the wild have allowed us to understand how natural selection works in considerable detail (see Question 11). While it has not been possible to create a completely new species, it has been possible to shift the traits exhibited by lineages within a species by a considerable amount—creating more or less nervous rats, or fruit flies with white instead of red eyes, or wingless lineages that happily breed true (i.e., produce wingless offspring generation after generation).

8. But isn't the theory of evolution just a theory?

It depends what you mean by "a theory." If you mean that a theory is something that hasn't been proved yet, then the answer is an emphatic "No." In science, we refer to these as *hypotheses*—they are suggestions for which we don't have any compelling evidence at the moment. The word "theory" means something very different in science. Theories are explanations, or models, that we have good reason to believe are true. Theories can come in two kinds: those that are true by definition and those that are true by reason of evidence.

The first kind are usually mathematical (or *deductive*) theories—something that must be true, given a set of assumptions about how the world is. We might have empirical evidence for those assumptions, or we simply might take them to be true because it is difficult to imagine any other way something could work—or because we just want to see what the consequences of their being true would be. Darwin's theory of evolution by natural selection is such a theory: its formulation as a set of three assumptions and a logical consequence (see Question 4) is true by definition (or deduction). The only possible issue of doubt is whether the three assumptions on which it rests are individually empirically true.

Theories of the second kind are simply empirical descriptions of the world based on summarizing many observations and are usually known as *inductive theories*. This kind of theory is always the workhorse of a science in its early days: you need to know what you have to explain, and to do that you need to make observations. Once you have done that, you may build a theory of the first (mathematical) kind to provide an explanation. Mendel's laws of (genetic) inheritance (see Question

22) were initially worked out empirically, and a set of very simple rules developed based on assumptions about the way traits are inherited. During the 1930s, these were combined with ideas from population biology to create a rich mathematical theory of evolution (the *modern synthesis*) (see Question 23). This has provided framework for research that has proved to be unusually productive in generating hypotheses to test.

There is an important distinction between theories that provide the overarching framework for our explanations and the specific subsidiary theories that we test that usually explore the mechanisms of that theory. Evolutionary biologists usually just assume that the theory of evolution is true and then devote their time to testing hypotheses about the way it works or the consequences we might expect it to have in the real world. Although it would be perfectly reasonable for scientists to assume that their framework theory is true without ever knowing whether this is in fact the case, in practice they usually have good reasons to assume that these framework theories are in fact true. It might seem odd, but assuming a framework theory is true by fiat is a perfectly reasonable thing to do in science: a theory simply provides a starting point for asking questions of the world. The important thing is that we keep testing the theory's predictions against empirical evidence and adjust the framework theory (or its assumptions) in the light of the results. In this way, science in principle can start at some completely random conceptual point and gradually, step by step, work its way around through this process of test-and-adjust to a theory that *is* a correct description of the world.

The formal mark of a theory is that it can be proved wrong. Darwin's theory could fail at any number of points, in particular its three fundamental assumptions, or axioms. If there was no variability among individuals in the extent to which they expressed a trait, or no evidence for the inheritance of those traits, or no evidence for differential reproduction, then Darwin's theory would fail. There are a number of other possibilities. Darwin's conception of evolution envisaged life having a single origin and then proceeding to diversify into ever more variable and complex forms as organisms encounter new environments and novel selection pressures. By implication, the story of life on earth should proceed from simple to complex. Broadly speaking, this is what we see (see Question 32).

A genuine challenge to Darwin's theory of evolution would be evidence for complex creatures before the appearance of simple multicelled organisms—or even at the same time as the simplest early organisms. One such example might be a Precambrian rabbit. This would imply the existence of something as complicated as a rabbit to be present in the fossil record at a time when the only other organisms were bacteria or simple multicelled creatures. Were such a rabbit ever to be found, it would cause immense difficulty for our current theories of evolution.

Generally speaking in science, the best way of proving—or disproving—a theory is to pit it against an alternative and see which makes the better predictions. The natural comparison for Darwin's theory is Lamarck's (see Question 2). In all tests so far, Lamarck's theory fails to outcompete Darwin's. Darwin's theory explains many things that were only subsequently discovered in the twentieth century—such as symbiosis (two species living together in close harmony) or the viral or bacterial origins of some components of our bodies (such as the mitochondria that exist inside each of our cells and provide the cell with its energy source) (see Question 36).

9. Then why did so many people not believe Darwin?

Darwin's theory of evolution, though widely supported by most scientists, ruffled the feathers of many laypeople, especially religious people who saw it as threatening the very sanctity of man's relationship to God (itself obviously a hangover from the Great Chain of Being) (see Question 2). Darwin's theory was perceived as implying that humans, so far from being next in godliness to God and his angels, were no better than apes. Strictly speaking, this was blaming Darwin for something that wasn't actually his fault. Over a century earlier, Linnaeus had classified humans with the great apes. Darwin merely provided the mechanism that might make this biologically plausible. Yet Darwin got the blame rather than Linnaeus—mainly because he mentioned evolution and Linnaeus did not!

Perhaps Darwin attracted such an adverse response because Linnaeus's classification of humans, chimpanzees, and orangutans in the same genus (as it happens, the genus *Homo*, the human genus, making the two apes honorary humans) still left open the possibility that humans were a special creation that God had chosen to nest among the apes. Linnaeus's classification was quite neutral about whether or not (or even how) evolution occurred. Even though the suggestion that anatomically similar species are likely to be related was implicitly built into his classification, Linnaeus made no explicit evolutionary claims in this respect: it could all have happened by special creation by the Creator just as it says in the Bible. Darwin's theory, on the other hand, ruled out any possibility of special creation. It implied that humans had evolved from some ape-like precursor through the same mechanisms of natural selection that have governed the evolution of all species. God's role seemed to be much reduced.

Not only did Darwin's theory challenge man's status in the universe, but for biblical literalists it also contradicted the claim made in the Bible that everything was created in one week—in October 4004 BC to be precise, according to Bishop Ussher's calculations (see Question 4). Darwin's mechanism of natural selection implied that evolutionary processes were slow and likely to take many millions of years to bring about change. It also implied that new species appeared piecemeal, emerging through natural selection from ancestral lineages from time to time over very long periods rather than in the ordered sequence described in the opening chapter of the book of Genesis, or even the Great Chain of Being. Since the Bible had been interpreted as the word of God, this seemed to challenge God's honesty or omniscience, or both. For fundamentalist Christians, this is tantamount to blasphemy. Of course, once again, this was not entirely Darwin's fault since it was the geologists who had decided that the good bishop's timescale was completely wrong. Yet, as before, Darwin got the blame because his theory added something about humans to the story.

The reasons fundamentalist Islam has since objected to Darwin's theory are quite similar. Islam considers the Koran to be the word of an omniscient God, dictated by him to the Prophet Mohammed in the seventh century AD. The difficulty this creates is that any new theory not included in the original text implies that God did not know about it and so cannot be omniscient: that is bad news, because in Islam anything that challenges the omniscience of God is blasphemy and punishable by death. Unluckily, evolutionary theory was not part of the philosophical consensus in the seventh century AD when Mohammed wrote his thoughts down, so it was never included in the Koran. And if there is one thing that is undeniably true of Islamic theologian and jurists, it is the fact that they are rigorously logical. It creates a bit of a problem.

10. Why is it important to ask why?

In many ways, the intrinsic complexity of the biological world is best summed up by what are known as *Tinbergen's four whys*. These were articulated in 1960 by Niko Tinbergen, one of the founding fathers of ethology (the naturalistic study of animal behavior) and co-winner of the 1973 Nobel Prize in Physiology or Medicine. Actually, three of the four "whys" were first stated by that great Greek philosopher-scientist Aristotle in his biology books, written around 350 BC, and were restated in the 1930s by Julian Huxley (one of the founding fathers of the modern synthesis) (see Question 23). Tinbergen added the fourth, and so usually gets the credit.

Tinbergen pointed out that when biologists ask why something is the case (for example, why do dogs bark?), they could be asking any of four completely different kinds of questions. These are questions about function (what benefit does barking have for the dog?), mechanisms (what caused the dog to bark on a particular occasion or what physiological machinery makes it possible for the dog to bark?), ontogeny (how do genes and the environment interact to give rise to an adult that barks from an embryo that doesn't?), and, lastly, phylogeny or history (what was the sequence of changes that occurred between an ancestral species that didn't bark at all to the contemporary species that does?). We might think of them as why, how, what, and when.

Tinbergen (and Aristotle!) made the important observation that each of these four "whys" are quite independent of each other. The answer to one has no effect at all on the answer to any of the other three. This allows us to investigate each one separately, even if we have no idea what the answers to any of the others might be. We can study behavior or ecology in the absence of knowledge of how animals develop such behavior, or the historical evolutionary processes that gave rise to it. By the same token, we can study a behavior's historical origins or developmental pathways without having to know why animals need their particular anatomical designs. Without this, the study of biology would be all but impossible.

To illustrate why, consider the case of how animals solve the problem of running fast (the functional explanation is that animals who run fast can escape predators or catch their prey more effectively)., Because four-legged animals move their legs at much the same frequency and cannot change this, running faster usually involves increasing the length of each stride. Different species achieve this by different mechanisms: cheetahs by evolving a very flexible spine that allows the spine to curve upward, enabling the front and back legs to stretch out farther on each stride; ostriches by evolving very long legs (allowing a longer reach on each stride); kangaroos by evolving the capacity to hop, thereby covering a longer distance on each stride. In other words, a particular function might be supported by several alternative mechanisms—or, conversely, a particular mechanism can sometimes support several different functions (see Question 13).

Perhaps the most important lesson is that we need to be careful not to confuse different kinds, or levels, of explanation. Functional explanations, for example, are usually concerned with the extent to which a trait allows an individual to contribute genes to future generations—although we might use a proxy for that such as how efficiently the animal forages, on the reasonable assumption that finding the food it needs will ultimately allow it to reproduce more effectively. Hence, suggesting that an animal forages because it is hungry (a motivational, or mechanistic, explanation) is not an alternative to suggesting that, ultimately, it forages in order to contribute more genes to the next generation. In fact, both explanations are needed. Genes cannot make animals behave, so natural selection has to work by setting up motivations as intermediates to make the organism do what needs to be done in order to achieve its evolutionary goal of maximizing its fitness (in effect, the number of genes contributed to future generations, or some proxy for this: see Question 25). (This is not, by the way, a circular argument: I could have phrased it more accurately, but it would have been even more complicated to read *and* much more boring. An important watchword in science: anything that is boring most probably isn't true or doesn't matter.)

Most of the disagreements over evolutionary explanations (and especially those between biologists and other disciplines) arise because each party is thinking in terms of a different kind of explanation (answering a different "why" question) but assume they are talking about the same thing. When zoologists ask why an animal behaves in a certain way, they often mean how a behaviors help the organism maximize its genetic fitness (a functional explanation), whereas a psychologist usually thinks in terms of the underlying motivations (it behaves in a certain way because it is hungry, or out of love, or from a parental instinct). Similarly, when zoologists ask how a behavior enables an animal to maximize its genetic fitness (the number of genes contributed to the next generation), they are thinking in terms of Mendelian genes (traits or characters), but other non-biological disciplines often mistake this for an explanation in terms of DNA (an ontogenetic explanation). Although biologists use the term "gene" for both, the word actually means quite different things in the two contexts.[5]

The four "whys" are important because they remind us how easy it is to make mistakes when considering complex biological explanations. This is because biology is a systemic discipline—biological phenomena are systems that involve many components locked together in a complex arrangement of causes, constraints, and consequences. We make egregious errors if we confuse these different components when trying to explain biological phenomena. This will, I hope, become abundantly clear as we consider particular examples later on.

2

EVOLUTION AND ADAPTATION

11. How do species adapt to their environments?

Darwin argued that species became increasingly adapted to their environments as variants (or what we would now call genetic mutants) that exhibited a slightly better fit to their environment survived better and were able to reproduce more successfully. The more successful types would gradually come to dominate in the population and so push the species' appearance inexorably toward their particular form as the less successful variants fail to reproduce and are weeded out.

We now know that this is exactly what happens. Peter and Rosemary Grant provided clear evidence of this in their 30-year study of Darwin's finches on the Galapagos Islands. They were able to demonstrate small but significant shifts in beak size within a population as food resources changed from one year to the next, making beaks of different size more, or less, efficient for obtaining seeds of different sizes and degrees of hardness. Darwin's finches provide a neat example of micro-evolution at work—the small, gradual changes in a trait that occur in response to changes in the environmental challenge.

Sometimes, however, large-scale adaptations can occur very fast. These often appear in the fossil record as an *adaptive radiation*—a period when a lineage undergoes rapid speciation over a very short geological time period. These occur most commonly when a species invades a new habitat where it has no competitors (a condition called *ecological release*). Under these circumstances, the population grows very quickly and then subdivides as the various populations start to compete ecologically with each other. One example of this is provided by the lemurs of Madagascar. It seems that around 40–50 million years ago, early prosimian primates similar to the galagos of mainland Africa drifted across the sea to Madagascar on vegetation rafts such as very large trees that had been uprooted and washed out to sea by major river floods (a not uncommon occurrence in this part of Africa even today). At that time, Madagascar was uninhabited aside from a few bird species. With no mammal competitors (and, more importantly, no predators), the colonists reproduced rapidly in the luxuriant conditions offered by the island and then diversified into many different species. Archaic humans (the ancestors of the Neanderthals) (see Question 62) probably became regular hunters soon after they first entered Europe from Africa for the same reason: Europe was well stocked with large herd animals (bison, horses, deer, mammoths) but few predator species, and these early humans were able to move into a predator niche in a way they had not been able to do in Africa where there were well-established guilds of small and large predators that long predated the appearance of the genus *Homo* (the genus to which we modern humans belong).

The best example of an adaptive radiation, however, is the rise of the mammals after the extinction of the dinosaurs at the end of the Cretaceous era 65 million years ago (see Question 47). The demise of the many dinosaur species over a relatively short period of time made a wide range of ecological niches available. The ancestral mammals (mainly small squirrel-like, tree-living creatures or badger-like ones that scurried about in the vegetation on the forest floor) rapidly diversified into a whole range of new lineages, including the ancestors of the primates, the insectivores, the carnivores, and the ungulates.

12. How do we recognize adaptations?

Conventionally, biologists identify adaptations in one of two ways. One is to show that a trait or character is anatomically or physiologically suited to the function it serves (that its mechanisms are well designed for its purpose); the other is to show that individuals that possess the trait in a more exaggerated way have higher fitness (in terms of either survival or reproductive success). The evolutionary biologist Theodosius Dobzhansky observed that the difference between these two approaches was that between being adapted and the process of adapting. The latter is a historical view of the active process of adaptation in action; the former is where that process has finally gotten us. These provide important tests of the underpinnings of evolution, so bear with me while I detail some of them. They provide the key to evolutionary theory's empirical success.

One of the best examples of the first approach is the eye. The eye's function is quite obviously to allow us to see the world around us in more detail and over a wider area than can be done by touch or scent. A detailed examination of its anatomical structure indicates that its design and the way it works is indeed appropriate to serve that function. It has an opening to allow light to enter, a mechanism (the lens) that focuses the light on the back of the eye, and a light-sensitive surface (the retina) at the back of the eye that responds to the light striking it from the lens. Stimulating the cells in the retina activates neurons (the optic nerve) that carry the signal to the brain, which then processes it and allows us to see. This is exactly how a camera works, and the proof of the pudding is, in many ways, that we have been able to build functional cameras using exactly this design.

A closer look suggests some more refined adaptations, at least in the vertebrate eye. One is a diaphragm (the iris) in the opening (the pupil), which adjusts the intensity of the light entering the eye. This prevents the retina from being burned out by excessively bright light, but also allows more light into the retina when light levels are low (e.g., at night), thereby increasing the amount of light stimulation falling on retinal cells. The lens itself is also well adapted to its particular function: it is composed of a material with a high refractive index that greatly reduces the blur of the image on the retina. Its shape (with thinner edges, like any magnifying glass) allows more refraction so that the light is concentrated on a smaller area of the retina, yielding a sharper image. This focusing is assisted by the vitreous fluid in the eye, which also has a higher refractive index than air, helping to focus the beams of light. The eye is a marvel of engineering design.

Eyes have evolved independently as many as 50 times across the animal kingdom, with many intermediate versions, from simple designs similar to pinhole cameras (such as those of planaria and the nautilus) to the compound eyes of many insects with their multiple lenses, and the equally sophisticated, but very different, eyes of vertebrates. Many of these can be seen as stages in the evolution of advanced vertebrate eyes caught midway to a more sophisticated form. These represent different lineages' solutions to the same environmental problem (how to see where you are going and how to avoid predators). As such, the eye provides some indirect evidence for the power of natural selection, since there is no reason why they should all converge on essentially the same solution if evolution was completely random (see Question 29).

More interestingly, all animals that have eyes in some form have the same genetic basis for eyes (perhaps because eyes are actually specialized outgrowths of the brain) and all of them use the same opsins (light-sensitive proteins) as light detectors in their retina. This may be an example of *convergent evolution* whereby unrelated species arrive at the same solution to a problem because there really is only one way of solving the problem physically or chemically. However, there are many differences between different lineages that reflect the specific demands of their lifestyles. Bees and many insects have eyes that are sensitive to ultraviolet light (which we cannot see) because many flowers display colors at this wavelength. Finding such flowers is, by and large, not of great interest to us, but it is a life-or-death matter to bees in their search for nectar. Similarly, the eyes of eagles are *much* sharper than ours, enabling them to spot prey at very great distances when flying. Our eyes, like those of all the monkeys and apes, are adapted more to close-up detection of fruits, other people's facial expressions, and predators (but only when they get close enough to be a nuisance).

There have been two distinct ways of undertaking the second approach to identifying adaptations. One is the approach that Darwin himself pioneered: using comparative data to determine how two traits correlate with each other across species. One such study looked at the difference in testis size as a function of mating system in primates to test the hypothesis that males in promiscuous mating systems need to be able to produce more sperm (and hence require bigger testes to do so) than males in monogamous mating systems. Since males face more competition from rivals, being able to swamp the female's reproductive tract with sperm is an advantage for a male in promiscuous mating systems because it increases the chances that it will be his sperm that fertilizes the egg, irrespective of how many other males have mated with the female. This leads to an escalating *evolutionary arms race* in which individuals (or species) steadily increase some trait in order to outcompete their rivals. Left to their own devices, an arms race would eventually result in males with giant weapons or testes. That this does not happen is due to the fact that all such developments are costly, and eventually the physiological or maneuverability costs begin to outweigh the reproductive benefits of the traits. Technically, this is known as *stabilizing selection*—whereby smaller and larger values of the trait are selected against, the one because animals are less competitive and the other because they incur too heavy a cost.

These kinds of analyses suffer from the disadvantage that they simply identify a correlation between two traits. Correlations do not allow us to infer causation (which trait is selecting for the other). However, there are now some very sophisticated statistical methods that allow the direction of causality to be tested in these kinds of analyses by determining the order in which traits change over the course of a taxon's evolutionary history. This is done by using data on contemporary species to reconstruct the most likely sequence of changes in traits over their common ancestry. These methods depend on knowing exactly how species are related to each other, of course, and therefore require sound methods for reconstructing phylogenetic histories (the evolutionary histories of different groups of species) (see Questions 30 and 44).

The other way of testing the process of adaptation, sometimes also known as the "baby counting" method, asks whether those individuals that have a particular trait in bigger or better form have higher fitness as a result. Do bigger animals in a population survive for longer than smaller ones, or, more importantly, do they produce more offspring? Peter and Rosemary Grant demonstrated this in their study of Darwin's finches: birds with heavier or shorter beaks (shorter beaks provide more power when biting) are able to crack open harder seeds when seeds are in short supply, such that they are more likely to survive through bad times and still reproduce (see Question 11). Similarly, in species of mammals like deer or some monkeys that have mating systems in which males fight for control over access to breeding females, do males who are more successful in defeating rivals sire more offspring—and do those males have bigger horns or canine teeth with which to fight?

In some cases, tests of this kind can be done experimentally. A nice example is provided by another bird study. It aimed to test whether the half-meter (20-in) long, rather ungainly tails of the male African widowbird were designed to advertise a male's quality to passing females and persuade them to mate with him (rather in the way a peacock's tail does). In an experimental study carried out in the wild, the Swedish biologist Malte Andersson caught males and artificially lengthened or shortened their tails. When they were released back to their display arenas, males who had their tails lengthened attracted many more females than males whose tails had not been altered, while males who had had their tails shortened attracted fewer—even though they were able to defend their mating territories just as well as other males. This study is also a nice demonstration of sexual selection (see Question 6) at work, and particularly of the fact that it is female preferences that are driving the evolution of the trait.

A particularly successful stream of research of this kind focused on the efficiency of foraging, known later as *optimal foraging theory*. It used experimental methods in both the laboratory and the field, as well as naturalistic observations, to test whether animals maximized energy intake as a function of the resources available to them. Its premise was that if natural selection had fine-tuned a species' decision-making abilities, animals would be able to balance the richness of individual food sources against their availability, preferring foods that maximized energy returns against the costs of procuring and processing them, switching from one food source to another just at the point where the first food source became uneconomical. It used the mathematics of optimization from microeconomics (itself borrowed from statistical physics) to make very precise predictions about animals' behavior, many of which were confirmed by the experimental results. In effect, it tested, very successfully, the adaptiveness of animals' decision-making abilities and hence how well natural selection had fine-tuned these capacities.

Very occasionally, modern molecular genetics offers us novel ways of testing adaptations. Until they died out as recently as 4,000 years ago, mammoths lived in the Siberian tundras that ran along the southern edge of the Arctic ice sheet. From time to time, one would die in a blizzard and become encased in an icy tomb, later exposed as a complete fully fleshed specimen as the ice melted in the later twentieth century. By extracting DNA from their carcasses, it has been possible to look at the differences between their DNA and that of their elephant cousins living at more equatorial latitudes. It turns out that their hemoglobin genes differ at four key positions. Those four genes make the difference between mammoth hemoglobin still being able to transport oxygen when atmospheric temperatures are near-zero and living elephant hemoglobin that would freeze solid under these conditions. This is what allowed mammoths to survive Siberian winters in ways that modern elephants could never hope to do.

13. How do new traits arise?

Although it is by no means impossible, it is unusual for new traits to arise completely de novo (from nothing). In most cases, they arise by adapting an existing trait for a new purpose. Here are two examples.

The ossicles are three tiny bones found in the inner ear of mammals. They link the eardrum (the thin sheet of skin in the outer ear that vibrates to sounds transmitted through the air) to the cochlea in the inner ear that converts these vibrations into neural signals to be sent to the brain, thereby allowing us to hear sounds. At just 3–5 mm long (less than quarter of an inch), these three tiny bones are the smallest in the human body. They derive from three of the jawbones of the reptile-like ancestors of the mammals. All reptiles have a lower jaw that consists of five bones on each side that are weakly fused together. Because reptiles often travel (think of snakes) or lie (think of lizards or crocodiles) with their jaws resting on the ground, they hear through their jaws: vibrations in the ground caused by animals moving about are picked up by the jawbones and transmitted to the brain as auditory (sound) signals.

When early mammals evolved out of this ancestral reptilian stock and became adapted to chewing plants, they needed teeth that were more robustly anchored—reptile teeth have rather shallow roots and are easily lost when put under pressure from chewing. This in turn required a jaw that was more robust and less prone to twisting. To solve this problem, the earliest mammals adapted the two bones at the front end of the jaw, fused them together, and strengthened them to hold permanent teeth. The three bones at the other (skull) end then became redundant, but their secondary auditory function predisposed them to form part of the auditory system as the ossicles once they had been miniaturized. Like our reptile ancestors, we still hear through our "jaws."

Another example of the way natural selection adapts traits is a hormonal one. The neuropeptide oxytocin has attracted a great deal of popular attention as the "love hormone" because it seems to be intimately involved in facilitating the trust and bonding associated with romantic relationships. In fact, this useful neurohormone is very ancient, having evolved in fishes to manage the body's water balance in an environment where it was necessary to prevent the body tissues from absorbing too much water. When some of these early fish eventually invaded dry land to give rise to the reptiles and amphibians, oxytocin continued to serve this same purpose—though now by keeping the body's fluid levels up despite the desiccating environment. Later, when the first mammals evolved, oxytocin acquired an additional important function: maintaining the body's fluid levels in the face of the drain placed on it by lactation. From here, it was adapted to facilitate maternal bonding to the infant so as to ensure continued willingness to provide milk and protection for the infant. And from here, it was adapted to facilitate pair bonds (the "romantic" relationship between the parents) in monogamously mating species so as to ensure there were some babies to lactate for.

In fact, there are often many ways of achieving the same end, so not all adaptations for the same purpose will look the same. The similar shape of fishes and marine mammals like dolphins both reflect adaptations to the demands of efficient swimming, despite beginning with bodies of very different shape and structure. Similarly, the wings of flying insects, birds, and bats are all designed to enable efficient flight, though in different ways that reflect the various species' anatomical origins. Colugos (or flying lemurs) and other gliding mammals represent a halfway house for this, in that by stretching a flap of skin between the arm and leg on each side of the body to provide a gliding platform, they are able to glide very effectively over distances of up to 100 meters.

14. Is mimicry an adaptation?

Mimicry refers to cases in which one species adapts its appearance to look like another. One could consider it an unusual example of convergent evolution (see Question 13). The most familiar cases, which have fascinated naturalists and biologists alike for more than 150 years, are those in which some perfectly palatable species looks like a species that is highly poisonous. This known as *Batesian mimicry*. Examples include many palatable butterfly species that look like the decidedly poisonous *Heliconius* butterflies. Providing the palatable mimic is less abundant than the poisonous model, a predator is likely to try the model first and be put off catching anything else that looks even vaguely like it. If the mimic becomes too common, the strategy won't work because a naïve predator will come across many mimics before it tries a poisonous one. More exotic examples include octopuses of the genus *Thermoctopus*, which, when approached by a predator, are able to adjust their body shape and color to resemble a sea snake or a lionfish—both of which are highly toxic. The plant known as the chameleon vine has the capacity to alter the shape and color of its own (very edible) leaves to resemble the less palatable ones of the particular host plant that it happens to be colonizing, thereby deflecting herbivores' attention from eating it.

An equally common form of mimicry is *Müllerian mimicry*, whereby two toxic species converge to resemble each other.

This allows them to exploit each other: if a naïve predator tries one and finds it distasteful, it will avoid the other as well. In many such cases, they also adopt a conspicuous coloration, such as red or orange, that makes them easily visible. Monarch and viceroy butterflies have such a relationship. Viceroy butterflies take this one step further by also having subspecies that resembles species of *Danaus* butterflies whose habit of consuming poisonous milkweed makes them highly toxic to most predators.

An unusual form of mimicry is automimicry, whereby one part of an organism's body looks like another part (often the tail looking like the head). This design distracts predators into attacking the tail rather than the more vulnerable head, making it possible to effect an escape by leaving its tail behind. The pygmy owl has eyespots at the back of its head, probably to mislead its own predators into thinking they have been spotted—most predators will not bother to attack a prey animal that has spotted it because, having lost the advantage of surprise, it is unlikely to be able to catch the prey. Caterpillars of the hawk moth also have eyespots on their terminal segments; when alarmed, the head retracts, leaving the large "eyes" on its rear end facing the predator. Thinking the prey must be too big to handle, the predator hastily backs off and goes in search of something it can handle more easily.

Perhaps the cheekiest form of mimicry is that adopted by some plants whose flowers are designed to look like the female sexual organs of their target prey. Known as pseudocopulation mimicry (or Pouyannian mimicry after its discoverer), it is particularly common in orchids whose flower parts often resemble the females of the bees and wasps that are the plants' main pollinators. It encourages the pollinators to go to them rather than to other species of plants, and so maximizes the likelihood that pollen will be transferred from one flower to another of the same species.

Camouflage, whereby an animal blends into its background to avoid being noticed by predators, is also a form of mimicry.

Some aphids mimic thorns or leaves so as to avoid detection by passing predators. Ambush predators such as chameleons, the praying mantis, and the devil scorpionfish all rely on camouflage in different ways to catch unwary prey that haven't noticed them. Many bottom-living flatfish like plaice and even the wobbegong shark flap their "wings" when they settle on the sandy seabed to scatter sand over themselves so that they blend in better.

By far the most famous example of camouflage as an adaptation is the peppered moth. Prior to 1811, it had only a pale form; however, after the onset of the coal-fired Industrial Revolution in Britain, large numbers of dark forms were found around the major northern industrial towns; by 1895, pale forms accounted for only 2% of these populations. In a classic field experiment during the 1950s, Bernard Kettlewell showed that birds were less likely to find the dark mutants on the soot-stained trees of the industrial north, whereas in the cleaner air of southwest England the darker moths stood out and were taken by the birds much more often than the pale moths.[1] It was predation pressure and detectability against a background that had resulted in the evolutionary shift, providing a convincing example of both camouflage and Darwinian adaptation.

15. What do vestigial organs tell us about evolution?

Vestigial organs are examples of body parts that have become dramatically reduced in some lineages to the point where they no longer serve any function. They provide some fascinating insights into, as well as compelling evidence for, evolution, since there is no way these could exist unless they had been inherited from ancestors for whom they had once performed a useful function. And now there they are, serving no function at all, and sometimes even getting in the way, a wistful memory of past history.

In some cases, the original organ was lost or greatly reduced in size because it was no longer needed when a lineage adopted a new style of living such as moving from the land back into the sea, as the whale family did. In many such cases, traces of the original organs can still be seen. Examples include the small, unattached leg bones still found buried inside the bodies of baleen whales, or the much-reduced wings of emus and ostriches that are now no longer capable of supporting flight. Other examples include the vestigial toes buried within in the horse's foot (a reminder of its four-toed ancestry) (see Question 7), the blind eyes of moles and subterranean fishes that live in the dark, and the tiny spurs that can be seen either side of the cloaca of pythons and boa constrictors that are the remnants of its pelvis.

Humans have a number of vestigial organs, such as the coccyx, or tailbone, that is buried in the flesh at the bottom of our spine (something we share with the other apes, of course) and the *plica semilunaris* (the remains of the nictitating membrane that forms the "third eyelid" in birds and reptiles whose function is to protect the eyeball without blocking vision). The coccyx can be a painful nuisance when it breaks in an awkward fall or, as is not uncommon, when women give birth. The appendix (the 9-cm [4-in] long sac at the junction of our small and large intestines) is thought to be what is left of the cecum that, in some primates, houses the bacteria used in the digestion of leaves—an explanation, incidentally, originally suggested by Darwin himself.

The phenomenon known as "goosebumps" is what remains of our ability to raise our body hair (as dogs and chimpanzees do when they raise their hackles when threatening): the little muscles around the base of hair follicles can still contract, causing the skin to pucker up, even though there is now no hair to erect. Many of the reflexes seen briefly in newborn babies have a similar origin: the grasping reflex that makes it possible for a newborn baby to carry its own weight by gripping onto an adult's two fingers (the grasping reflex) and the Moro, or startle, reflex (whereby a newborn grabs out with its hands if it is dropped gently onto its back) both derive from the fact that all Old World monkeys and apes carry their infants under their bellies, with the infant holding onto the mother's fur with a grip of iron. These reflexes disappear naturally when the baby is a few weeks old.

Our wisdom teeth (the third, rather small, molar at the back of each jaw quadrant) are another example of a vestigial organ: they are much reduced in size, and some populations don't even have them (e.g., Mexican Indians, where their complete absence is associated with two specific genes). If they emerge at all, they do so only during the late teens or early 20s (hence their supposed association with the attainment of wisdom)—an adaptation in many mammals to prolong the life of the grinding teeth when relying of a diet of coarse vegetation that needs a lot of chewing. (Elephants take this to the extreme: due to their very coarse diet, they only ever have one molar in each jaw quadrant at a time, with each tooth being ejected in turn once it has worn down so as to allow the next one to come through.) Their loss reflects an evolutionary change in the human diet late in our evolutionary history that placed less emphasis on the heavy grinding of food. Because of this, wisdom teeth cause us many problems: they easily become impacted against the neighboring tooth as they start to surface, leading to considerable pain, as well as elevated risk of caries if food rots around them.

Not all vestigial organs are nonfunctional. In some cases, they may have been adapted (technically *exapted*) to new purposes, often reshaped and repositioned. The best-known example, mentioned earlier are (see Question 13), are the ossicles (the three tiny ear bones buried deep in our inner ear) that derive from remodeled ancestral reptilian jawbones. The wings of penguins are another example: these can no longer be used for flying, but they have been *very* successfully adapted for swimming. The gill slits (or pharyngeal arches) that briefly appear in the embryos of both chickens and humans very early during fetal development may be another example: they quickly become transformed into the jaw and neck. Because their position and form resemble those found in fish embryos that give rise to the gills that allow fish to "breathe" in water, it has been suggested that they may be part of our biological inheritance dating back to our fishy ancestors.

Since true vestigial organs have no function, it is very difficult to see how they could exist had they not been inherited from an ancestor that had once had a use for them. They are also powerful evidence against the suggestion that the world was created as we see it. No omniscient creator starting with a clean slate would ever have dreamed of including these largely pointless and sometimes inconvenient organs in any species he or she was designing—least of all the ones that end up causing us problems in the way our appendix and wisdom teeth sometimes do.[2]

16. How fast does evolution happen?

The speed of evolution depends on the pressure of selection. That, in turn, often depends on large-scale changes in the earth's climate (see Question 17). The extent to which animals and plants are affected by climate change, however, depends on where they live. Oceanic habitats, and especially those at great depths, are relatively less affected by changes in surface climate. For this reason, species that live in the ocean depths tend to undergo less evolutionary change. Modern sharks, for example, are almost identical in shape and design to fossil sharks from 100 million years ago. Indeed, the oldest members of the shark family date back 400 million years, long before either plants or animals had colonized the land surfaces.

In other cases, where organisms are exposed to more intense levels of selection, evolution can be quite fast. A very rough estimate is that it will take a new trait about 1,000 generations to spread through a species and become the norm. For humans, with a long generation length (the time between your birth and the birth of your own offspring) of about 25 years, this is equivalent to around 25,000 years; but for a virus that can have many hundreds of generations in a day, that can be as little as a week or two.

When the selection pressure is strong enough, however, traits can evolve quite rapidly. Polar bears and the brown bear are closely related, the polar bear having evolved from a population of brown bears that became isolated in eastern Siberia at the height of the last glaciation a mere 20,000 years ago. The striking differences in appearance between these two species and the polar bear's ability to swim and to withstand the intense cold of the Arctic thus evolved over a relatively short period during which the population must have been under very intense selection pressure.

Another example is provided by the ability that some humans have to digest milk as an adult. In most mammals, including humans, the ability to break down and digest the lactose sugars in milk is normally switched off when babies are weaned—usually around four years of age in pre-agricultural populations. As a result, both children and adults of most human races suffer from lactose intolerance: drinking raw milk causes vomiting and diarrhea, and even death if consumed in quantity. The capacity to tolerate raw milk as an adult is unique to Caucasians (Europeans) and some cattle-keeping peoples such as the Fulani of West Africa and the Maasai and related tribes of East Africa (who may all have Mediterranean origins). It must have originated in the ancestral population of Europeans when they first domesticated cattle a mere 10,500 years ago. The need to be able to drink milk as adults became essential when these populations invaded Europe, where the low light levels characteristic of Europe made it difficult for the skin to synthesize vitamin D, which in turn affects calcium absorption (see Question 68). Milk provided readily available sources of both—providing you could tolerate the lactose in it. Lactose tolerance represents a specific adaptation to a completely different problem faced by human populations trying to live at high latitudes—a nice example of how interconnected biological processes can sometimes be.

Although evolution can occur quite quickly when selection is intense, most selection pressures are in fact quite modest. Estimates of the selective advantage for a range of traits under selection in the real world (in effect, how much faster a new mutant trait contributes genes to the next generation compared to the standard trait) are typically in the region of a very modest 5–10%. At that rate, it will take a great many generations for a newly evolved trait to spread through a species and become the norm.

17. How does climate influence evolutionary change?

Evolution occurs when circumstances change sufficiently to challenge animals' abilities to survive and reproduce successfully, when competitors or predators disappear, or when new opportunities open up for animals migrating into a new habitat. Smaller-scale changes can occur as a result of relatively modest changes in the quality of the environment or its vegetation on a year-to-year basis. Large-scale evolutionary change is often caused by substantial changes in the climate that occur over a period of decades rather than millennia. These are often associated with events like ice ages or rapid climate warming that have a dramatic impact both on animals' ability to maintain thermoregulation and on the quality and quantity of their food supplies. Darwin himself recognized that understanding the major geological and climatic patterns in the earth's past history was the key to understanding evolution.

Broadly speaking, the earth's climate has been getting progressively cooler over the last 500 million years. For long periods in the past, the northern landmasses were dominated by warm shallow seas, with tropical conditions prevailing even at the latitude of London. From around 2.5 million years ago, the earth descended into a period of prolonged cooling with the onset of a series of ice ages. Alternating colder (glacial) and warmer (interglacial) periods of increasing intensity culminated in the current Ice Age, which began around 500,000 years ago and reached its peak around 30,000 years ago in a series of cycles of colder and warmer intervals of around 100,000 years in combined length.

While it is not completely clear what caused the ice ages or why they cycle in the way they do, geologists think that they reflect various combinations of movements in the earth's tectonic plates (and hence the positions of the continents, and the effect these have on the flow of air masses and oceanic currents), the buildup and dissipation of methane and other greenhouse gases, the Milankovitch cycles (the rather regular wobble in the earth's orbit round the sun; see next paragraph), variations in the moon's distance from the earth, the occasional impact of very large meteors, and volcanic activity.

The Milankovitch cycles (named after the Serbian geophysicist, Milutin Milanković, who discovered them in the 1920s) have three key components, each with its own cycle length: the eccentricity, or shape, of the earth's orbit round the sun (varying between closer and farther away over a period of 100,000 years); the obliquity, or tilt, of the earth's axis (which rotates over a period of 41,000 years); and the axial precession, which affects the direction in which the poles point and is caused by tidal forces exerted on the earth by the sun and moon in about equal proportions (over a period of exactly 25,771 years). In addition, some extra cycles due to the gravitational pull of the giant planets Jupiter and Saturn have also been detected more recently, though their effects are much weaker than the three main ones. Each has very specific effects on the earth's climate because of the way they affect both the amount of radiation that reaches the earth from the sun and the evenness with which it does so across the hemispheres. When all these cycles coincide (as they do every 400,000 years), the earth experiences more extreme weather conditions.

Then there are the rather more mysterious reversals of the earth's magnetic field. Reversed polarity (when the magnetic south pole becomes the magnetic north pole) occurs at completely unpredictable intervals and can last for periods ranging between a few hundred and tens, or even hundreds, of thousands of years. Geologists have not been able to determine why they occur when they do, but major comet collisions are one possible explanation. It is, however, thought that changes in polarity may be associated with extinction events, possibly because they are associated with active volcanic activity that throws vast quantities of dust into the atmosphere that blocks out the sun, causing a partial nuclear winter (see Question 46).

Climate change can sometimes be extraordinarily dramatic and very rapid. At the end of the Younger Dryas climate event around 10,000 years ago, average global temperatures rose by 7°C (12°F) in as little as 50 years. Current concerns about global warming of a few degrees almost pale into insignificance beside this catastrophic climatic event. The Younger Dryas itself was a brief 2,000-year return to ice age conditions in the northern hemisphere that seems to have been triggered by the collapse of the ice wall that dammed the great Canadian glacial Lake Agassiz. Huge quantities of cold water had been released into the North Atlantic, causing sea levels to rise by about 13.5 m (44 feet) over a 300-year period; it blocked the Gulf Stream (otherwise known as the North Atlantic conveyer) that brings warm water from the Caribbean up into the North Atlantic to give Britain and northwest Europe its mild winters. This caused northern hemisphere temperatures to plummet by as much as 10°C (18°F) over a period of about 1,000 years. Glacial conditions and tundra returned to the northern landmasses. Then once the cold northern waters had mingled with the warmer waters from the south, global warming kicked in again, marking the end of the Ice Age.

The important implication for evolution is that terrestrial climates never stand still: they are always changing, sometimes slowly, sometimes extremely fast. As well as affecting how hot or cold, and how humid, a place is, climate has a major impact on the types and quality of vegetation that will grow at a particular site. Between 11,000 and 5,000 years ago, the cooler, wetter climate resulted in what is now the Sahara being quite lush, with elephant, giraffe, and other woodland animals in abundance. Climate warming over the last thousand years or so has resulted in its progressive desertification.

Between them, climate and vegetation are the major determinants of how suitable a location is for a particular species. A species that is perfectly adapted to one set of climatic conditions can suddenly find itself completely ill adapted a few thousand years later. If the pace of climate change is slow, vegetational and thermal zones will move slowly, and a species may be able to track them. If they change fast, a species can find itself in the wrong place at the wrong time, and is likely to go extinct very quickly (see Question 45).

18. Is perfection the inevitable outcome of evolution?

One common misunderstanding of evolution is that it leads inexorably to organisms that are ever better designed. But Darwinian evolution is not a process of perfection. In contrast to the Lamarckian and Great Chain of Being theories of evolution, natural selection does not inexorably drive species up some kind of ladder of perfection. Rather, natural selection is, to borrow a term from economics, a process that leads to satisficing (just being "good enough for now"). In other words, so long as you survive and do better than your rivals, that is good enough. An antelope does not have to become the fastest animal on earth; it simply has to be faster than the lions that try to catch it. By the same token, lions and other predators do not have to be able to run marathons at top speed; nor do they have to be able to catch every prey animal. They just have to be fast enough to catch sufficient prey animals to survive.

Second, and perhaps even more importantly, evolution is a Heath Robinson process: it always starts with a particular organism and just tinkers with its design—making small changes here or tweaking a trait there. Every species has a history, and that history is the baggage that constrains what it can do. In addition, because an organism is a complex system of different parts that have been adjusted to each other, most changes are more likely to lead to the death of the individual than to any sudden glorious new future. Even small changes are likely to have adverse consequences for some other aspect of the organism's biology and threaten its survival—a reminder that organisms are complexly integrated systems whose components have been carefully honed by natural selection to work well together.

Suppose, for example, that I wanted to reach up higher while feeding. The advantage of doing so is that it provides me with access to food sources at tree heights that other herbivores cannot reach. One obvious way to do this is exactly what the giraffe has done, namely to increase the length of my neck. That can be done quite easily just by tweaking the gene that controls the size of my neck vertebrae so as to allow them to grow a little larger: giraffes have the same number of neck vertebrae as the rest of us, but each of them is much longer so that it ends up with a longer neck. However, that also increases the amount of muscle needed to hold the neck upright, which in turn means I'll need to eat more food to fuel these muscles. It also means I'll need a larger heart to pump blood that extra distance vertically against gravity. And at the same time, it means that I'll need better valves in the veins to prevent all the blood rushing back down from the brain under the force of gravity. I will also need some mechanism to prevent the blood flooding into the brain and bursting it when I lie down. Our simple problem quickly becomes a series of additional problems that also need to be solved, which may explain why not too many species have opted for this particular strategy. In a word, climbing around in trees the way monkeys do might actually have been the easier evolutionary option.

19. Why do some biological features seem to be poorly designed?

Natural selection can work only with the material it is given. It simply tries to improve on this when circumstances change. The constraints imposed by the way a particular part of the body is designed, or the way any given trait interfaces with other parts of the system, may limit how much it can be changed to fulfill a new function. This commonly results in adaptations that are less than perfect—or, at least, less perfect than if one were to design an organism from scratch.

Perhaps the most famous example of this is provided by the vertebrate eye. The nerves in the retina that carry the signals from the photoreceptor cells in the retina to the visual processing areas in the brain are gathered together at the back of the eye to form the visual nerve to the brain. The most sensible way to do this would be for the individual nerves from each retinal cell to emerge from the back and then be bundled together. In fact, they emerge on the *front* side of the retina and so have to pass back through the retina before going to the brain. This creates a small area where there are no light-sensitive cells—the "blind spot" where we cannot see anything. To compensate for this, the eye jitters from side to side so that the light that would have fallen on the blind spot occasionally falls on the vision cells around it. The brain then has to adjust for this movement to create an integrated picture based on averaging across the different inputs it receives over time. This is hardly going to win the prize at the ideal design exhibition.

What makes this doubly puzzling is that it doesn't have to be this way. The octopus's eye is designed the other way around: the nerves from the photoreceptors lie behind them and pass out of the eye as a bundle without having to pass through the retina, so they have no blind spot. This may be just one of those 50:50 cases where a mechanism could have developed either way: it did so one way in the octopus family and the other in the ancestor of the vertebrates.

Another famous example of a less-than-perfect adaptation is that in vertebrates (but not invertebrates), the sensory neurons cross over so that sensations from the left side of the body are processed in the right side of the brain, while those from the right side are processed in the left side of the brain. This rather odd—and hardly ideal—arrangement occurs because at an early stage of development (but after the basic neural structures have been laid down), the embryo twists in such a way that body and the brain become reversed. Although this has been known for a very long time, no one knows why it happens this way.

A more familiar everyday example of the inefficiency of adaptation is the human back. Our ancestors' decision to walk bipedally (on two legs) rather than quadrupedally (on four) like all the other monkeys and apes has imposed significant stresses on the lower vertebrae in the back because it now has to carry the full weight of the head and torso. Of course, we could easily have solved this problem by evolving massive lower vertebrae with lips to protect the discs and stop them from "slipping" (the main cause of lower back pain). But that would have dramatically reduced our spinal flexibility and hence adversely affected our ability to run—and, especially, to throw spears (during which the whole upper body twists). Instead, we seem to have opted to accept that some individuals will suffer back problems as the price for the rest of us being able to run and jump—a reminder, perhaps, that evolution focuses on the average individual, not on perfection for all.

In an important sense, the evidence for poor design is evidence against a divine designer. We wouldn't really expect an omniscient designer to have done such a bad job quite so often. We would have expected him (or her) to have anticipated future problems and done something about them right from the start. In contrast, mistakes of this kind are exactly what you would expect in Darwin's theory, since all changes (mutations) build on the organism's existing state and do not anticipate future needs or circumstances.

20. Why do we sometimes fall prey to genuinely destructive addictions?

Humans fall prey to a large number of diseases that are known as the "diseases of civilization" because they rarely occur in hunter-gatherers (the lifestyle in which we have spent 99% of our evolutionary history as a species) but are very common among people who live in the postindustrial world. These include obesity, diabetes, heart disease, and drug and alcohol addiction. We might see these as design faults where natural selection has done an incomplete job. In fact, they have more to do with the fact that modern humans in postindustrial societies are living under conditions that are completely novel and have been in existence for only a few thousand years at most.

Some of these diseases are consequences of selection for a perfectly sensible preference for sugars and carbohydrates (the main sources of energy). Since these are often in short supply in nature, a motivation to gorge on them when they are available is beneficial for traditional hunter-gatherers, even though it leads to obesity and diabetes when this is done in the conditions of plenty provided by the modern world.

Alcohol offers a particularly interesting example. Alcohol is, in fact, a poison, but we and our African great ape cousins (the chimpanzees and gorillas) evolved the capacity to exploit it around 10 million years ago. Alcohols are produced in most rotting fruits by natural yeasts that settle on them from the air. The yeasts convert the sugars in ripe fruit into alcohol (as a waste product of their fermentation process). In the wild, overripe fruits typically contain 1–4% natural alcohol, which is a valuable source of energy *if* it can be converted back into sugars. The African great apes evolved an early genetic mutation in two key enzymes that allow them to do this. The alcohol dehydrogenase enzyme ADH converts the alcohol into acetaldehyde, which is then converted into acetic acid by the aldehyde dehydrogenase enzyme ALDH; the acetic acid can then enter the Krebs cycle, where its chemical energy is extracted by the same mechanism that converts sugar into energy. Everything depends on the speed at which these two enzymes work in tandem. If ADH is too slow, alcohol builds up and we get drunk; if ALDH works too slowly, there is a buildup of acetaldehyde, which is highly toxic. No other monkeys (or any of the Asian apes) are able to handle alcohol as well as humans and the African great apes.

The reason for this unusual mutation seems to be that when the climate took a sudden downturn during the Miocene era around 10 million years ago, the great tropical forests contracted into small relict patches. Apes, which had been the dominant primates for the previous 10 million years, were suddenly thrust into greater competition with monkeys. Apes in general (including ourselves) cannot digest unripe fruit, which are full of tannins and other phenolics designed to prevent herbivores from eating fruits before the seeds are ready to germinate (see Question 52). Because monkeys had evolved the capacity to detoxify unripe fruits (probably as an earlier adaptation to allow them to eat leaves), they could outcompete the apes by being able to eat the fruits before the apes could. The result was a wave of ape extinctions that reduced the number of ape species to just 10% of what they had been. However, the forest floors were littered with overripe fruit that had fallen from the trees (or been dropped by carelessly feeding monkeys). A mutation that allowed apes to handle the alcohol in the fruits offered a massive advantage because it gave them access to an important food source that monkeys could not exploit (both because they rarely came to the ground and because, even when they did, they were ill adapted to coping with the alcohol). The Asian apes (gibbons and orangutans) didn't acquire this mutation because they had branched off much earlier during the Miocene and fruit-eating monkeys didn't get established in Asia until much later.

The medical profession has tended to see alcohol use as a disease. Indeed, recent large-scale epidemiological studies confirm that excessive use of alcohol increases the risk of diseases such as heart conditions, cancer, diabetes, and dementia. As part of our general dietary strategy as apes, however, it is a natural feature of our evolutionary history, and one that has come to be used in all human societies as part of the mechanism for social bonding. And in this latter respect, it seems to have served a particularly important adaptive function: the consumption of alcohol not only triggers the main pharmacological mechanism used in social bonding (see Question 87) but also forms part of the rituals used in social bonding worldwide.

As with almost all nutrients (and even essential chemicals such as oxygen and water, or trace elements like iron and phosphorus), the relationship between the quantity ingested and its benefit is almost always an inverted-U shape: the more you have, the better—up to a point, after which it becomes increasingly poisonous. Like everything we consume, too much of a good thing invariably turns out to be a bad thing, and it has been humans' capacity to industrialize everything (and hence make it stronger, more digestible, cheaper, and more easily available) that has been responsible for creating problems. Once again, evolution could not anticipate future conditions.

3

EVOLUTION AND GENETICS

**21. Why was the discovery of genetics so important
for our understanding of evolution?**

Despite the fact that Darwin's theory of evolution was widely
acclaimed by his scientific contemporaries, it had a central
weakness that left it open to criticism. Neither Darwin nor
anyone else had any real understanding of the mechanism of
inheritance that was needed to place his theory on a sound
biological footing. Even though he—and all the plant and animal breeders of his day—knew perfectly well that traits were
passed from parents to offspring, they did not understand how
this was done or why some traits were inherited more reliably
than others. Some of the early reviews of *The Origin of Species*
had criticized Darwin on exactly this point.

Darwin never solved this problem, either to his own sat-
isfaction or to anyone else's. Indeed, in struggling to find
a solution, he was led to propose that each part of the body
contributed "gemmules" to the offspring that allowed it to reproduce each particular trait (a theory he termed *pangenesis*)
and that these mixed in equal proportions in the offspring (his
theory of *blending inheritance*). What he did not seem to appreciate was that this would have had the opposite effect to the
one he wanted. If parents' traits were mixed in equal measures at conception, the offspring would be an average of the

two parents; if this was repeated generation after generation, everyone would eventually end up being identical, and there would be little or no variation left in the population for natural selection to work on. The process of speciation would eventually grind to a halt.

The final straw came in 1892, a decade after Darwin's death, when the great German physiologist August Weismann published his theory of the germ plasm. Weismann argued that the relationship between the body (or soma, as he called it) and the germ plasm (the elements contributed by each parent to the embryo at conception) was a one-way process: germ plasm (or genes, as we would now understand this) could influence the soma, but the soma could not influence the germ plasm. (We would now refer to these as the *phenotype* and the *genotype*, respectively.) This principle became known as Weismann's central dogma. It meant that Lamarck's theory of the inheritance of acquired characteristics (see Question 3) was completely impossible: Lamarck's theory required the soma to be able to change the germ plasm so that repeated use could influence the acquisition of traits. Although Weismann himself supported Darwin against Lamarck, many people, especially those with other axes to grind, interpreted this at the time as evidence against Darwin's theory. In fact, many went so far as to argue that embryology and genetics explained evolution, and Darwin's theory was unnecessary. They seemed unaware that they were confusing two of Tinbergen's four whys (Darwinian function with ontogeny in the form of genetics). Nonetheless, because of this, Darwin's theory of evolution fell out of favor.

Ironically, the answer Darwin needed was already there. Indeed, Darwin may even have known about it but not appreciated its significance. Between 1856 and 1863, the Augustinian friar Gregor Mendel spent many—no doubt happy—hours in his monastery gardens at Brno in Silesia (now part of the Czech Republic) breeding pea plants in an attempt to understand how their traits were inherited. After testing some

28,000 plants (being a monk, he evidently had plenty of time on his hands), he managed to work out the basic principles (see Question 22). Unfortunately for science, in 1868 Mendel was appointed abbot of the monastery, and his administrative duties meant that he never had time to return to his experiments or even to publish his results. Then, horror of all scientific horrors, when Mendel died in 1884, his successor as abbot burned all of his papers. We know about Mendel's discoveries only because of a brief summary he published in a local scientific society journal just before he became abbot (a copy of which was in Darwin's library).

So when the Dutch botanist Hugo de Vries published the results of his own experiments on inheritance mechanisms in the evening primrose in 1900, some 15 years after Mendel's death, he was surprised to be criticized for having overlooked Mendel's work. De Vries eventually conceded Mendel's posthumous claim to priority, and, perhaps for that reason, it is Mendel who gets the credit for discovering the laws of inheritance rather than de Vries, whose name has been all but forgotten. Nonetheless, de Vries's scientific contributions live on: he gave us the word *gene* and developed the concept of a *mutation* (the idea that genes can change spontaneously into a new form), thereby laying the groundwork for the later development of modern genetics (see Question 24).

22. What are Mendel's laws of inheritance?

Mendel's discoveries are sometimes seen as an accidental discovery by the man in charge of the monastery garden. Nothing could be further from the truth. In fact, Mendel was well embedded in the scientific world of the Austrian Empire, having been a student (of philosophy and physics) at both the University of Olomouc (in Silesia) and the University of Vienna itself. At Olomouc, he had been inspired by Johann Nestler, whose principal research interests had been the inheritance of variation in animals. When Mendel was given permission to pursue his scientific interests back at his monastery, he chose to study variation in peas, which he could easily breed in the monastery's extensive gardens. He was not in charge of the gardens; he was merely allowed to use them for his experiments.

Gradually, over a long, painstaking series of experiments, Mendel figured out that each parent plant must pass on to their offspring a "factor" for each trait (we would now refer to these as *alleles*, or genes), so that the offspring received two factors for each trait. He recognized that these factors bred true: when the two parents had different traits (green versus yellow fruits, for example), the offspring did not exhibit a mixture of these traits, but typically looked wholly like one or the other parent. More importantly, looking at all the offspring produced by a single pair of parents of different type (those with green versus yellow fruits, for example), he discovered that they exhibited a surprisingly constant ratio of the two traits: one-quarter of the offspring resembled one parent (they had yellow fruits) and three-quarters resembled the other parent (and had green fruits), a ratio that remained consistent across many thousands of experimental crosses and generations.

Mendel reasoned that if each parent possessed two factors for a given trait and only one was passed on at conception, then the possible combinations in their offspring would follow a simple statistical rule if these segregated at random during the process of *meiosis* to produce gametes, or sex cells, that had only half of each parent's genes. (We now know that this happens through the segregation of the chromosomes on which all these "factors" lie, so that each gamete has only half of the parent's chromosomes.) When these pair up with the other parent's factors during fertilization, there are four possible ways the two sets of two parents' factors can pair up, and, as each combination is equally likely, each combination will account for a quarter of all the offspring. If both parents were *Aa* (where *A* is the factor for green fruit and *a* for yellow fruit), each would pass on the *A* factor half the time and the *a* version the other half of the time, and their offspring would have the combinations *AA, Aa, aA,* and *aa* in equal frequencies. This would yield 25% *AA,* 50% *Aa,* and 25% *aa.* If one of the two factors contributed by the parents was suppressed by the other (Mendel's concept of recessives and dominants), such that, say, *A* was dominant over *a,* then any *Aa* genotype would always appear as an *A* phenotype in the offspring. In such a case, a simple arithmetic rule would predict a 3:1 ratio of in favor of the dominant factor: the *AA, Aa,* and *aA* genotypes would all appear as an *A* phenotype and only the *aa* variant would exhibit trait *a.*

Two familiar human examples of simple Mendelian inheritance are hair color and eye color, both of which are inherited as single genes, with dark colors dominant over light colors. This means that the offspring of an *AA* (dark) homozygote (some with two copies of the same gene) and an *aa* (light) homozygote will all be *Aa,* and therefore dark-haired (and brown-eyed)—even though one parent was blond (and blue-eyed). But in the next generation, if the *Aa* (dark) offspring mate with an *aa* homozygote, they will produce three blond and one dark offspring. It's all a matter of simple logic.

A related, but even more fascinating, phenomenon was discovered in the 1980s: *genomic imprinting.* In contrast to a recessive allele being suppressed as in conventional Mendelian genetics, genomic imprinting involves the genes for a particular trait from one parent being completely switched off (or "imprinted") no matter what, so that the trait is always determined only by the genes from the other parent. About 150 genes are known to behave like this. For example, your limbic system (the brain's emotion center) is determined only by your father's gene(s), whatever they are, and your neocortex (the brain's "thinking" center, and the site of all conscious activity) only by your mother's gene(s). (I resist any temptation for an I-told-you-so comment, but feel free to make it anyway.) Another example is the IGF2 (insulin-like growth factor 2) gene, which is expressed only from the gene inherited from the father. In effect, the genes seem to "know" where they have come from.

For most traits, the alleles inherited from the two parents can compensate if one of them is defective, but this obviously cannot happen with imprinted genes. In humans, at least two well-known genetic disorders (Prader-Willi and Angelman syndromes) appear to be a consequence of genomic imprinting, the one due to a paternal effect (the active but defective gene is the one inherited from the father), the other to a maternal effect. Both give rise to developmental disorders, with significant social and mental disabilities. The gene DIRAS3, which appears to suppress the growth of tumors (especially in the ovaries and breasts), is also maternally imprinted. In cases where the defective gene is inherited from the mother, individuals are at significantly elevated risk of cancer. The reasons why this mechanism has evolved are intriguing, and I will return to it in Question 78.

23. What is the "modern synthesis"?

After Mendel's laws of inheritance had been rediscovered, a number of geneticists and evolutionary theorists during the 1930s and 1940s rescued Darwin's theory of evolution from the near-oblivion into which it had sunk by developing a mathematical theory of the evolutionary process that combined Mendel's laws with Darwin's insights. It was the title of Julian Huxley's 1942 book *Evolution: The Modern Synthesis* that provided the name for the new combined theory.

In effect, this merged a Mendel-derived genetic theory of microevolution (how the genetic structure of a population can change through time as the result of selection) with Darwin's original macroevolutionary theory of species change (how species evolve to adapt to different environments)—principles 2 and 3 of the Darwinian formula (see Question 4). The important thing was that the microevolutionary part was formed by a mathematically sophisticated deductive theory (predictions about the future composition of the population followed from a set of axioms and assumptions about the pattern of inheritance and the rate of selection), as opposed to Darwin's decidedly inductive theory of species evolution (derived from a more traditional compilation of examples to generate generalizations).

Armed with Weismann's "central dogma" (genes influence the organism's traits, but not the reverse) (see Question 21) as a core assumption and Mendel's semi-mathematical principles as its engine, the synthesis showed how populations of animals would evolve over time. The evolution of entirely new species simply required a big enough change in the environment (so increasing the strength of selection) and sufficient time for a suitable mutation to occur (so as to produce a trait that better suited the particular demands of the new environment).

The modern synthesis has, over the past 75 years or so since its creation, proved remarkably powerful in its ability to predict how the genetic structure of a population changes under the impact of natural selection and sexual selection, or indeed its absence (so-called *neutral selection*) (see Question 29).

In the final quarter of the twentieth century, modern synthesis also laid the groundwork for incisive new developments in evolutionary theory, such as the concepts of *inclusive fitness* and *kin selection* (see Question 26), and *stabilizing selection* (whereby anatomical or physiological constraints can maintain a trait within a limited range of variation over many thousands of generations because increases or decreases in the trait are deleterious, causing those who inherit these mutations to be reproductively less successful). It also gave rise to new fields like *island biogeography* (which explores how the number of species is related to the size and isolation of geographical regions like oceanic islands, oases in the middle of deserts, or even isolated mountains).

24. How did the discovery of DNA change our understanding of the genetic mechanism in evolution?

DNA, or deoxyribonucleic acid, was discovered by the Swiss biologist Friedrich Miescher in 1869 when he found that the nucleus of the human white blood cell contained protein-like strands, which he called *nucleins* (or what we would now call *nucleic acid*). Half a century later, the Russian biochemist Phoebus Levene discovered that these strands of nucleic acid consisted of a set of phosphate-sugar-base combinations. Then in 1944, Oswald Avery and his team at Rockefeller University, New York, demonstrated that the hereditary units of genes consisted of DNA. The final piece of the jigsaw came in 1953 when Cambridge biologists James Watson and Bernard Crick proposed, with the aid of crystallography photographs of DNA taken by Maurice Wilkins and Rosalind Franklin in London, that the strands of DNA formed a double helix along which were strung a series of four bases (thymine, adenine, guanine, and cytosine) connected across the two strands by hydrogen bonds. The four bases were always paired off in the same way, thymine with adenine and guanine with cytosine. That model, originally worked out with pieces of cardboard on a desk, has, with very minor adjustments, remained unchanged for more than half a century.

The helical form was what allowed the strands of DNA (or chromosomes) to replicate themselves. During cell division, the helix unwinds and each half rebuilds its opposite side from the cellular fluid as the bases attract their opposites. Most of the time, this process is amazingly reliable in reproducing daughter cells that have identical chromosomes to the parent cell. However, from time to time, a mismatch can occur—a base pair gets lost, or segments of DNA might break loose and be inserted elsewhere or the wrong way around. Such errors typically affect only one in every 100 million bases, and so are quite rare. In many cases, these errors disrupt the chemical balance of the cell and result in cell death, so that the process is self-correcting—although DNA strands turn out to have a great deal of redundancy, so that if one segment doesn't do what it is supposed to do, another can often step in. However, from time to time, the new combination works well enough and survives as a mutation that produces a new variant in a trait. Thus, the new understanding of how cell replication worked turned out to explain mutations as well.

It was very quickly worked out that the base pairs formed sets of three (now known as a *codon*) along the line of the chromosome, each of which codes for a different amino acid that in turn codes for a particular protein used in the process of building new cells. A simple "start" or "stop" instruction codon regulates how long a particular process of cell division is allowed to repeat itself, and this provides a very simple mechanism for changing an organ's size. In effect, building new cells involves reading the blocks of three base pairs along the chromosome from a "start" codon until a "stop" codon is reached.

This turns out to be a remarkably simple and elegant solution for the mechanism of inheritance, and one that does not require unusual conditions or processes. The triplet structure limits the number of amino acids needed to 64, though in most cases only around 20 of these are actually used by cells. From the point of view of evolutionary theory, however, the cracking of the genetic code turned out to have an additional benefit: it revealed that all species (except some of the most primitive archaebacteria) use exactly the same genetic code, implying that they evolved from a common origin, just as Darwin conjectured.

25. What is genetic fitness?

Darwin's theory of evolution by natural selection consists of three axioms (or principles) and a logical conclusion (see Question 4). The third of these provides the fulcrum that makes all this possible: individuals with a particular version of a trait reproduce more successfully. Evolution occurs because, in the long run, natural selection will always seek to maximize an organism's ability to contribute copies of its particular genes to future generations. That is simply a consequence of the three axioms being true (that inherited variants in a trait will contribute differentially to the next generation). The contribution the individual makes to the next generation is its *genetic fitness*.[1] Technically, fitness is a property of a trait or gene, and not of an individual, but biologists often refer to the fitness of the individual as a shorthand. It is crucial, however, to be clear exactly what this entails: failure to appreciate what is involved has been the cause of many misunderstandings.

Until the modern synthesis was developed, biologists commonly interpreted fitness in terms of survival (hence the phrase "the survival of the fittest"—coined, incidentally, by the political philosopher Herbert Spencer, and not by Darwin). However, successful reproduction and longevity (i.e., survival) are not necessarily the same thing. This was pointed out in the 1960s by English ornithologist and founding father of population ecology, David Lack. In a seminal series of field studies of songbirds, Lack showed that simply laying a lot of eggs did not necessarily maximize the number of chicks that a pair fledged. If food or time is in short supply, parents who lay too many eggs will overstretch their ability to feed their chicks; as a result, many—or even most—will die of starvation (in effect, Malthus's principle in action in the natural world) (see Question 4). If they have no surviving offspring, the pair will fail to contribute their genes to the following generation and their fitness will be zero. The pairs that do best are the ones that lay one more egg than the average number successfully reared in their local area: the extra egg gives them a buffer against accidental losses and some extra capacity *if* it turns out to be a better-than-average year in terms of food availability. This is now known as *Lack's principle*, and it reminds us that there are two separate components to fitness (survival and reproduction) and that these two can be traded against each other (see Question 59).

In the 1960s, W. D. Hamilton pointed out that if two individuals share a particular gene because they are related by descent from a common ancestor, then that gene can be propagated in the next generation through either individual. That being so, an individual might, in some circumstances, do better to help its relative reproduce than by trying to reproduce itself. Fitness thus consists of two components: the number of grandchildren that an individual produces and the number of *extra* grandchildren produced by *each* of its relatives as a direct result of the assistance provided to them by the individual concerned (taking into account the probability that they share the gene—in essence, how closely related they are by descent from a common ancestor).

Hamilton showed that this can be summed up in a very simple equation, now known as *Hamilton's rule*: you should be altruistic toward someone whenever the extra offspring they get through your helping them (devalued by the degree of relatedness between the two of you) is greater than the number of offspring that you lose by helping them. This is the formula:

$$rB > C$$

where r is the relatedness (the probability of sharing a gene), B the number of extra offspring gained by the relative, and C the number of offspring you forgo. He termed this modified definition of fitness *inclusive fitness*.[2]

Hamilton's insight is a reminder of three important things. First, when we calculate the fitness of a trait, we must determine not just the benefit provided by the trait, but also the cost the individual has to pay in order to generate that payoff. This is true even for simple, conventional fitness. Second, we have to compare the net benefit (benefit minus cost) with what it might have gained by pursuing any number of alternative options (the *opportunity cost*, or what economists, in a rare moment of unintended humor, used to call the *regret*—your regret at not doing something). In other words, we have first to ask whether the net benefit of doing something (the benefit minus the costs) is greater than zero; and then, second, harking back to Lack's principle, is *this* net benefit greater than the net benefit generated by any alternative strategy? If the costs are too high, a trait won't be favored by natural selection even if it offers considerable benefits. Finally, it reminds us that every individual is embedded in a community that ultimately extends to all life on earth, and that everything an individual does has consequences for the fitness of every other individual in that community. Because the probability that two individuals share a gene falls off very quickly as their relatedness declines, in practice the impact on fitness is very limited for individuals who are less closely related than cousins and so can mostly be ignored. Nonetheless, it is a valuable reminder that everything we do has consequences for others, and that this in turn has consequences for us (see Question 73).

26. Are genes really selfish?

The term "selfish gene" was popularized by Richard Dawkins in his 1976 book of that name. It is simply a metaphor for the processes of natural selection: genes acting under natural selection behave *as if* they were selfish. The genes are not themselves behaving selfishly. They are no more conscious of what they are doing than any other chemicals can be said to be conscious. The genes themselves don't really do anything in particular (other than code for the trait they represent). Natural selection is a blind process that results in some genes being differentially represented in future generations and others not. But, because we are so deeply attuned to thinking about a human world in which individuals have intentions, it is psychologically much easier for us to understand goal-directed processes like evolution if we think of genes as being concerned to maximize the frequency with which they are represented in future generations. It looks teleological, but it isn't: it is just a manner of speech.

Of course, there are all sorts of reasons why this might not always be true. One is that multicellular organisms (in other words, most of the life forms more advanced than viruses and bacteria) have a vested interest in the effectiveness with which the other genes that make up the organism they are part of are propagated. A multicellular organism is a composite of many different genes that, in effect, cooperate to produce an individual that can reproduce effectively. Any gene that goes all out for its own interests at the expense of the interests of all the other genes will not only be the author of its own extinction, it will also drive all the other genes associated with it to extinction too. In effect, it has to cooperate and be "mindful" of the other genes' requirements.

Another classic example is offered by pathogens. If they are so virulent that they kill off their host, not only will all the host's genes die out, but the destructive pathogen itself will also die out. What often happens in these cases is that the less virulent variants, or mutants, are the ones that are able to reproduce best—because they do not kill their hosts—and so they contribute most to future generations. In other words, the gene pool of the virus comes to resemble the less virulent forms rather than the original more virulent one. This phenomenon of reduced virulence as a pathogen reaches some kind of consensus, or stable equilibrium, with its host is extremely common. Many of the viruses that have inserted themselves into our DNA have probably arisen through this process. Many of the viruses that infect us with diseases likewise evolve reduced virulence with time.

A more serious consideration is that offered by many social species whereby the survival and reproductive success of the individual, and hence its constituent genes, depend on the part played by the other individuals in the social group. This might involve cooperation in hunting or simply ganging together as a defense against predation (see Question 81). In effect, it is a kind of implicit social contract, in which we all agree to stick to the rules because that way we will all do better. Individuals who do not play their part undermine the social contract and may thereby cause everyone to do less well than they might otherwise have done—and so drive them to extinction. It makes these kinds of phenomena unstable, but if they can be made to work everyone usually ends up better off (see Question 73).

These examples notwithstanding, when thinking about how we would expect natural selection to work, the assumption that genes will act in their own best interests works well for most practical purposes. It makes very clear predictions as to how a gene, or at least its host, should behave, all else equal. Then we can ask whether what we see is what we predict. As always in science, it is the occasions when our theories fail to predict what we see that are most useful since they identify circumstances where we don't know all the relevant components. This should make us reconsider our assumptions about how the world works and ask what we might, in our ignorance, have left out of the equation. This is the central core of the scientific method.

27. But how do animals know who their kin are?

The key to Hamilton's concept of inclusive fitness is obviously kinship (the degree of relatedness between individuals). But to make the calculation inherent in Hamilton's rule, an animal has to know who it is related to *and* to what degree. We humans can tell each other who is related to whom, but how on earth do animals do this? In fact, it turns out to be quite easy. They use a number of very simple proxies—ones that we humans also use.

One of these is whether or not you grew up in the same nest (or at least saw each other often while you were growing up). If you grew up in the same nest, it is more than likely that you are related. This even works with us. To be sure, we have adopted and stepsiblings who muddy the waters. However, these are actually quite rare by comparison with true relatives, so that if you grew up together, there is a very decent chance that you are in fact genetic relatives. In one of our studies, the cue that best predicted people's willingness to bear pain to earn someone else a cash reward was how often they had seen them during the first 10 years of their life. This correlated remarkably well with their actual relatedness.

Another mechanism for identifying relatives is physical appearance. Close relatives are more likely to look like each other simply because they have more of the same genes that determine physical appearance. For rather obvious reasons, this underpins the way we choose mates from the same species rather than trying to mate with anything that moves. Physical similarity is also a surprisingly good cue of much closer degrees of relatedness. In hamadryas baboons, male brotherhoods form clans that try to monopolize females, and male clan members in the same band (the social grouping that forages and sleeps together) can often be distinguished by their facial similarity.

Facial similarity and sometimes even similarity in other physical traits are cues we also use. In one study, Lisa DeBruine found that people were more generous in economic games when playing against a stranger represented on the computer screen by a photograph whose face had been manipulated to look more like their own. Indeed, the one thing that visitors to newborn babies do is comment on whom the baby looks like most. The mother's side of the family, in particular, will often emphasize how much the child resembles the father. For obvious reasons, mothers always know whose baby it is, but fathers can never ever be 100% sure. So, since the father needs to believe that the baby is his in order to be willing to invest in it (otherwise he runs into the problem of genetic altruism— investing in someone else's baby), it seems that his in-laws (in particular) work very hard to persuade him that the baby really is his. If you don't believe me, just listen carefully the next time someone you know has a baby. It works *every* time.

A third mechanism is smell. A great many species rely on smell to identify relatives. In part, this is because animals that live together (especially in nests) have a family smell. But, in fact, smell is a very good indicator of genetic relatedness because your natural smell is determined by the same genes (the MHC, or major histocompatibility complex, genes) that determine your immune system. Even tadpoles can recognize kin by their smell. Despite the fact that the sense of smell is much poorer in primates than in almost all other mammals, even humans are surprisingly good at recognizing individuals by their smell. And here's something else to watch out for: the way women family members, in particular, will pick up a new-born baby and hold it close to their face and take a surreptitious sniff. (It happens very quickly, so pay attention!) It's as though they are checking out who the baby really belongs to, though they usually excuse it by saying they love the smell of new babies. It is, by the way, an old primate habit: monkeys and apes do it too. Another example is the Maori *hongi*—the greeting usually misnamed by westerners as rubbing noses. In fact, they are simply putting their noses next to each other and breathing in deeply to exchange *ha* (the breath of life)—in other words, your smell.

Humans, of course, also use language to say who is related to whom. Anthropologists frequently complain that human kinship naming systems bear little or no resemblance to biological kinship, partly because some societies refer to female relatives of the mother as "mother" and partly because strangers and visitors are invariably given kinship status so as to place them in the community's family structure. There are, in fact, only six major kinship naming systems in the 6,000 or so languages that currently exist, and broadly speaking they all predict biological kinship much better than would be expected if they were completely random. More importantly, in an inspired but little-appreciated analysis, the evolutionary biologist Austin Hughes showed that all of these are essentially variations on how to identify relatedness in societies that differ in paternity certainty (how sure a man can be that he is the father of the offspring produced by his wife).[3]

In short, there are many different mechanisms that animals (including humans) use to identify close relatives. None of these mechanisms is perfect, of course. But then evolution doesn't require everything to be perfect—it merely requires that it works well more often than not (providing the costs aren't too great). Using several indices to triangulate the effect, of course, reduces the chances of mistakes, which may well explain why we use so many of them.

28. Are there other kinds of biological transmission besides DNA?

Most of the genetic material involved in heredity is contained in the chromosomes that make up the nucleus of each cell. A normal human cell has 23 pairs of chromosomes (including the X and Y sex chromosomes), half from your father and half from your mother, making 46 chromosomes in all. Between them, these contain around six billion base pairs (or two billion codons). However, it turns out that in humans, only around 20,000 of these codons (about 1.5%) are functional (that is, code for a protein), about the same number as in a fruit fly or a roundworm. The rest seem to do nothing in particular, which led to them being labeled *junk DNA*.

In fact, these noncoding regions consist of a variety of elements, some of which do important work in the less overtly visible background. Some code for RNA (ribonucleic acid) that provides the interface between the DNA itself and the proteins that DNA codes for, and some are regulatory sequences or repeats of particular codons. Others, however, are *pseudogenes* (codons produced by gene duplication that no longer generate proteins because they have acquired mutations that render them inactive): there are about 13,000 of these in the human genome. In humans, the genes for smell receptors have an unusually large number of pseudogenes (around 60%, compared to just 20% in mice), which probably explains our relatively poor sense of smell compared to most other mammals. Regulatory genes can be especially important (they control when a functional gene is activated) and make up around 8%

of the human genome. Many of these are conserved across taxonomic groups and are often used for estimating how closely related two species are (see Question 30).

Around 50% of the human genome consists of random repeats of a codon (known as tandem repeats as they stack up like echoes along the chromosome). These can be highly variable between individuals and are also often used in testing for genealogical relatedness or for forensic purposes. Some 44% of our genome consists of mobile elements (sometimes known as *jumping genes*) that can detach themselves and reinsert themselves elsewhere. Many of these appear to be retroviruses that have inserted themselves into our genome at some time in our deep evolutionary past after invading the body of a host. Most of these are now functionally neutral, mainly because those that behaved disruptively killed their hosts and so were the authors of their own extinction (see Questions 26 and 35).

An important class of nongenetic mechanisms of inheritance are those known as "maternal effects"—or "cytoplasmic effects" because they are usually contained in the nongenetic material in the cytoplasm of the egg. In sexually reproducing species (see Question 56), the sex that provides the sperm (or its equivalent) supplies only its nuclear DNA, whereas the sex that provides the egg contributes both nuclear material and the cytoplasm that surround the nucleus, along with its mitochondria (see Question 36). As a result, the mother can sometimes affect the phenotype of her offspring by contributing free-floating messenger RNA, proteins, or even hormones. Maternal effects have even been proposed as an explanation for childhood obesity: deficits in maternal cellular metabolic control have been shown to alter fetal pancreatic, fat cell, and muscle cell development, all of which can affect obesity.

Other important mechanisms for transmission may be purely behavioral. Parental behavior can have an obviously beneficial effect on an offspring's chances of growing up and reproducing successfully, and this may be especially important for social species where social skills need to be transmitted from parent to offspring by copying (cultural inheritance) (see Question 91). Another form of nongenetic inheritance may be *niche construction*, whereby a species alters the environment in which it lives so as to facilitate its own survival or that of its offspring. Examples include the dams made by beavers and the mounds of termites. But perhaps the single most important form of niche construction is the formation of social groups in intensely social species like monkeys and apes, and of course ourselves: these turn out to have a bigger effect on offspring survival than almost anything else (see Question 81). The inheritance of property and wealth in humans is also a form of nongenetic inheritance, since wealth inherited from parents allows offspring to be placed at a social advantage in most societies, thereby giving them a reproductive head start.

29. Can evolution happen in the absence of selection?

Evolution simply means change, and as such it does not need to involve the directed change that natural selection brings about. The process involved in such cases is often called *genetic drift*, mainly because the genetic makeup of the population was originally thought of as drifting at random rather than being pushed in a particular direction. Genetic drift depends on two key things: the background rate of mutation and the absence of selection. This gave rise to the *neutral theory of molecular evolution* developed in the late 1960s by the Japanese geneticist Motoo Kimura. This argued that most of the mutations at the molecular level had no effect on fitness (i.e., were neutral in their impact) and would simply accumulate in the species' gene pool over long periods of time, subject to accidental eliminations.

Mutations are a natural consequence of the process of genetic replication whereby errors of copying arise as new copies of a gene are produced during cell division (see Questions 21 and 24). The errors can come from several sources. Segments of DNA can become detached and reinserted the wrong way around, or in the wrong place, on a chromosome. Or a mistake may occur in the base pair as the DNA helix strand assembles its complementary half. Given the way genes are "read" (in sequential sets of three bases along the chromosome) during the process of building an organism, any slight change in the order or identity of base pairs will result in entirely new codons, and hence different proteins being manufactured (see Question 24).

This might seem careless of evolution, but in fact it is part of the design. In theory, it is perfectly possible for natural selection to create a copying process that is 100% accurate. However, that would mean that there would be no variation between individuals on which natural selection could act. And if that were so, there would be no evolution. So, paradoxically, natural selection makes the system less than perfect. Of course, too much error would also be disadvantageous, so we might expect the system to end up as a compromise—a balance between too much and too little accuracy in copying (another example of *balancing selection*). There is, however, one source of mutation that is beyond even the capacity of natural selection to deal with, and this is the effect of cosmic rays from the sun. Cosmic rays hitting the body commonly cause mutations (which is one reason why too much sunbathing is bad for you: damage to the DNA can cause skin cancers).

All else equal, new mutations will gradually accumulate in a population unless they are selected against, thereby increasing the variance in the population over time. Because closely related individuals will usually live near each other, it is then likely that different populations of the same species will have a slightly different genetic makeup. When population size is small, the chances of some of these mutations being passed on can vary considerably simply due to the likelihood of either an offspring inheriting one variant rather than another or of a particular population surviving. As a result, the genetic compositions of two neighboring populations can gradually diverge. This provides us with a natural index of how similar two species are genetically, which in turn provides us with the basis for the so-called *molecular clock*.

30. What is the molecular clock?

The molecular clock was first postulated in the 1960s, and is arguably the single most useful consequence to have emerged from molecular genetics. It is a direct consequence of the neutral theory of molecular evolution (see Question 29). By being able to compare the DNA of two individuals at the level of individual bases, we can determine the number of point mutations (i.e., individual base pairs) by which any two species (or even individuals) differ.

The key to the molecular clock is the probability that any one base pair will flip into another per generation. Only noncoding DNA (so-called *silent mutations* because they have no effect on the organism's appearance, or phenotype) should be counted since these are the product only of the (more or less) constant natural rate of mutation. By counting up the number of base pairs at which two individuals differ, we have a direct measure of how closely related they are genetically. From this, we can count the number of generations needed to produce that number of differences. The calculations are quite complicated because we need to factor in the likelihood that a base pair might flip from one state to another and then back again later if the time interval is long enough.

Some of the other factors that can throw off the absolute timing of the clock include changes in generation time (neutral mutations will accumulate more slowly in species with slow reproductive rates, and hence long generation times) and demographic bottlenecks when only a small subsample of the population survives to become the next generation (which can speed it up). Bottlenecks can result from either a small population budding off from the main species when it invades a new habitat (the *founder effect*) (see Question 43) or from major population crashes. In each case, only a small proportion of the variation in the species' gene pool survives, often causing the species to lurch dramatically toward one extreme of its phenotype. Generally speaking, these errors are quite modest, and the estimates for divergence times are still orders of magnitude better than anything else we have (aside, perhaps, from radiometric dating of rocks and materials—which relies on the very precise decay rates of atomic processes). In any case, we can set statistical error limits around the date estimates given by the molecular clock to tell us how accurate the clock is in any given case.

Perhaps the most spectacular example of the clock's use has been in calculating the date for the origin of our own species, anatomically modern humans (*Homo sapiens*). This has involved determining how many base pair differences there are in the DNA of a large sample of living individuals from as many human populations as possible, and then using the clock to calculate how long ago they all shared a common ancestor (known as the *convergence date*). Separate estimates have been calculated using mitochondrial DNA (mtDNA, which is inherited only down the female line) (see Question 36) and Y-chromosome DNA (which is inherited only down the male line) (see Question 57). For mtDNA, the rate of change (or substitution of one base pair by another) is about 0.02 per base per million years; this is 5–10 times higher than is the case for nuclear DNA, thus providing a rather more precise clock. The estimated age for our species is 100,000–160,000 years according to various estimates using the mtDNA clock, and 120,000–156,000 for the Y-chromosome clock. In other words, our species is only about 150,000 years old. Given that the average lifespan of the species in our lineage back to the common ancestor with the other great apes is about half a million years (the Neanderthals, for example, survived for well over 400,000 years) (see Question 62), we are mere evolutionary teenagers by comparison.

4

EVOLUTION OF LIFE

31. Can we say how life on earth started?

The origins of life on earth must predate the earliest fossils, and presumably must be simpler in form than even those earliest fossils. It is very unlikely that we will ever know what these life forms were, or when they lived, simply because they will have been too small and fragile to leave any record. Our earliest evidence of life dates from around 3.5 billion years ago—about a billion years after the earth collided with its smaller twin planet Theia (our moon was formed out of the ejected materials). These ancient life forms were cyanobacteria (a single-celled photosynthesizing microbe) that left fossils in the form of stromatolites (layered mounds of rock produced by their colonies) such as those found along the coast of Western Australia.

We can, however, make some informed guesses of how life on earth came about. First, there are only two options: either life evolved here on earth (and might be a unique event in the universe) or it arrived here from somewhere else, perhaps on a comet or meteor or, in the form of very simple organic molecules, as cosmic dust. If the latter is true, that's interesting but simply pushes the question back to somewhere else: how did life evolve wherever it did in the universe? One suggestion is that the conditions for the formation of these molecules might have existed in the disk of planetary dust that surrounded the young sun before the planets condensed out of it; if so, life might well have evolved around many other stars too, and we may be far from unique.

The fact that all forms of life on earth today share the same subset of 355 genes and the same cytochrome-c (see Question 7) suggests a common origin, with chemical chains formed from these components as the earliest kind of organic life. Of course, it is quite possible that this wasn't the first organic matter to evolve: there may have been several false starts that died out, our living world here on earth as we have it being the only one that succeeded in the sense that all living matter is now descended from it.

All life here on earth depends on four key chemicals: lipids (fats), carbohydrates (sugars and cellulose), amino acids (protein metabolism), and nucleic acids (the self-replicating molecules). Any explanation for the origins of life needs to explain how we got all four of these. What is clear is that the conditions required for life are quite narrowly defined: a watery environment that is not cold enough for water to freeze, plus the wide availability of the chemical elements (nitrogen, carbon, ammonia) that are the basis of DNA. More importantly, these molecules had to be able to form in the absence of oxygen, since the planet became oxygenated only much later (around 2.5 billion years ago) (see Question 32).

One early suggestion (the "primordial soup hypothesis") was that lightning strikes in the earth's turbulent early atmosphere provided the energy for these chemicals to fuse together to make simple organic molecules that later grouped together to form RNA (ribonucleic acid, which typically takes the form of a single strand of nucleic acid that, in the form of messenger mRNA, interfaces between the DNA of the genetic code and the production of the proteins that make up cells) and then later DNA chains. The famous Miller-Urey experiment in 1952 attempted to demonstrate this in the laboratory, using electric sparks to simulate lightning and an atmosphere of water, methane, ammonia, and hydrogen that was assumed to replicate the earth's early atmosphere: the result was amino acids of the kind that could form DNA.

An alternative suggestion, however, is that these early organic molecules evolved in the very hot, watery environments of deep-sea hydrothermal vents where hot gases bubble up through the earth's crust, creating a watery gaseous environment at 100–150°C (212–300°F) (i.e., above the boiling point of water). Even today, these deep-sea environments are populated by hyperthermophilic bacteria that seem able to survive in very hot environments rich in methane, ammonia, carbon dioxide, and hydrogen sulfide, most of which are toxic for the majority of living organisms.

Recently, astrophysicists discovered that the moon is moving away from earth at a rate of about four centimeters (1.5 in) per year. A simple calculation tells us that it is now about 17 times further away from us than it was when it first formed. A moon so close would have had dramatic effects on the oceans: the planet would have experienced tides of around three kilometers (nearly two miles) twice a day, compared to an average closer to 10 meters (33 feet) today. These would have scoured out the land surface over huge areas and deposited mineral-rich soils on the seabed, providing the chemicals (especially iron) needed to evolve living molecules in hot-water vents. In other words, life on earth may have been an accident of the size and distance at which our moon happened to come into existence, and so could be unique.

32. How did life on earth as we know it evolve from these very primitive conditions?

The very earliest forms of life that appeared around 3.5 or so billion years ago were, in effect, naked strands of RNA floating about in a hot, nutrient-rich soup. Once these RNA strands, and the double-helix DNA strands that they gave rise to, could reproduce themselves effectively, they diversified rapidly. But still, this was largely an anaerobic world operating in an environment all but devoid of oxygen. Today, oxygen constitutes around 20% of the earth's atmosphere and is the crucial fuel on which all multicelled life—that's you and me—depends. But it was not always so. The earth's geological evolutionary history seems to have involved a series of major phase transitions in which the conditions on the planet underwent a dramatic change in a relatively short period of time.

The first of these is known as the Great Oxygenation Event, which occurred around 2.5 billion years ago (so about a billion years after life first appeared). It seems that cyanobacteria in the oceans evolved the capacity to use the sun's rays to create energy by photosynthesis: this produces oxygen when it converts carbon dioxide and water into glucose (as an energy source), with free oxygen as a waste product. This gave these bacteria a massive advantage, allowing them to form dense living carpets on or just below the sea surface, much like the algal blooms we sometimes see today. We can be sure there must have been vast numbers of them because to produce enough oxygen to change the planet's atmosphere in the way they did would have required the release of huge quantities of oxygen: small quantities of free oxygen are quickly absorbed by the iron molecules in water (to create iron ore, which itself is now plentiful in the earth's crust because of that), and would not have resulted in an oxygenated atmosphere.

An oxygenated environment, whether in the surface layers of the oceans or in the atmosphere itself, is highly toxic for anaerobic bacteria. More importantly, the free oxygen would have neutralized the methane (a greenhouse gas) in the atmosphere and created an ozone layer that reduced the amount of sunlight hitting the earth. This eventually resulted in a glaciation that was so intense and all-encompassing that the earth effectively became a giant snowball (*snowball earth*), resulting in a major die-off of life forms. Under the resulting intense selection pressure, the capacity to exploit oxygen as a reducing agent for chemical processes emerged in some bacterial lineages, resulting in an equilibrium atmospheric composition and an entirely new trajectory for the evolution of life. Explaining how this happened is, however, still a major challenge. Nonetheless, the oxygen-rich atmosphere triggered a massive radiation (see Question 11) of new life forms.

The second major phase transition in the history of life on earth was the invasion of land surfaces by plants around 470 million (half a billion) years ago. This required the evolution of decay-resistant cell walls to prevent desiccation of the delicate living matter in the much drier conditions on land. With the benefit of super-powerful electron microscopes, we can see their spores preserved in rocks even though they are as small as 30 micrometers (less than 1/100th of a millimeter) in diameter. The speed with which plants evolved and colonized the land surface was remarkable: prior to 470 million years ago, there is little or no fossil evidence for land plants, but after that date they are everywhere. Biologists are still unsure why this happened so quickly.

By 370 million years ago, plants with woody stems and leaves had appeared, probably as a direct result of the plants competing among themselves for access to sunlight. Plants evolved stems to try and get above their competitors so they could receive the full glare of the sun on their leaves for photosynthesis, much as trees do in tropical forests today (which is why they are so tall, and the ground vegetation beneath them is so sparse). This gave rise to the forests that produced our coal seams when they eventually died. (Grasses, in contrast, were latecomers: they didn't appear until 40 million years ago as the climate started to become drier with a higher concentration of carbon dioxide in the atmosphere.) Plants helped to stabilize the wind-blown sandy soils that predominated on land, both through their root systems and by creating peaty soils that enriched the land surface for other plants. These new land environments created opportunities into which whole new groups of animals (and, in particular, the reptiles and amphibians) radiated and evolved, exploiting these plants as food sources.

33. Why do bacteria and viruses evolve so fast?

Life forms on earth are divided into two major groups, the prokaryotes and the eukaryotes. The eukaryotes have cells surrounded by a membrane and genetic material in a cell nucleus, whereas prokaryotes are typically single cells, sometimes just naked strands of RNA or DNA. Bacteria make up most of the prokaryotes. Having been the first form of life on earth, they continue to be a major component of living matter. A single gram (a third of an ounce) of soil contains around 40 million bacteria, and a liter (one quart) of water contains around 100 million (which explains why the famous Lamarckian experiment demonstrating spontaneous creation of life seemed to worked so convincingly) (see Question 2). Bacteria perform all manner of essential and valuable functions without which life on earth could not exist. This includes fixing nitrogen in the soil for plants to use and decomposing dead bodies so as to return their constituent chemicals to the soil.

Viruses seem to have evolved later, possibly from free-floating plasmids (small pieces of DNA that can move between cells in an organism's body) or directly from bacteria. They are essentially tiny segments of RNA or DNA contained within a protein coat, and can normally reproduce only when they are inside the cell of another organism (plant or animal). They were discovered in 1898 by the Dutch biologist Martin Beijerinck in the form of the tobacco mosaic virus (which causes mottling on the leaves and badly damages crops). Viruses' capacity to move easily between organisms (even organisms of different species: aphids, for example, acquire them when feeding on the sap of one plant and then transfer them to another plant), and their ability to insert themselves into organisms' cells (even into their nuclear DNA) (see Question 35), means that they have been responsible for a great deal of horizontal transmission of genetic material between species. It is this capacity to insert their own DNA into another organism's DNA that makes possible genetic engineering of crops and potentially sophisticated medical treatments.

While both bacteria and viruses can perform useful functions (or at worst be neutral), they are also some of the main purveyors of the diseases that cause us so many problems. The big advantage that both bacteria and viruses have over more conventional animals and plants is that they can replicate very rapidly by simple cell division (mitosis). Most animals and plants have a natural generation time because sexual reproduction (see Question 56) means that two parents have to meet up and mate in some way in order to reproduce. Bacteria and viruses can just divide and reproduce so long as the conditions are suitable. As a result, they can explode numerically very quickly—which is essentially what happens when we fall ill: the bacteria or viruses reproduce much faster than the body's immune system can destroy them.

Because they can reproduce so fast, bacteria and viruses can try out many hundreds of thousands of mutations over a very short period of time: those that don't work die out, but those that work can get through the sieve created by both the body's immune system and our medical knowledge. In this way, they can stay one jump ahead of our attempts to kill them off. This is why entirely new winter flu variants appear each year despite the fact that we have managed to develop anti-flu vaccines for last year's versions. It is also why MRSA-type bugs evolve that cannot be treated by antibiotics even though these worked just fine on previous infections. This is the evolutionary process at its simplest: new genetic variants are produced when an organism is placed under intense selection by a change of environment (in this case, the invention of antibiotics capable of killing off large numbers of bacteria) (see Question 40).

34. Why did multicellular organisms evolve?

Multicellular organisms (those that consist of many cells living together as a single organism in the way that plants and animals do) could evolve only after a solution was found to how an organism could be produced from a single germ cell. The problem arises from the fact that most such organisms consist of a variety of cells, each performing a different function and often having a very different anatomy as a result.

Multicellularity has in fact evolved some 46 times in the eukaryotes, as well as in some prokaryotes (cyanobacteria are one example). Complex multicellularity that involves many different specialized cell types, however, has evolved only in six major groups of eukaryote organisms: brown algae, red algae, green algae, fungi, land plants, and animals. Animals represent the most extreme case since they have 100–150 different cell types, compared to just 10 or 20 in plants and fungi. Many plants, the vertebrates in general, and some arthropods have carried this one stage further by evolving a separation between somatic cells (the ones that make up the body), which are sterile (i.e., cannot reproduce), and germ cells (gametes), which can reproduce. Somatic cells simply live through their life cycle and die, leaving no descendants. Only germ cells have the capacity to replicate and (after fertilization) create a whole new organism. In effect, somatic cells sacrifice themselves to enable their germ cell sisters to reproduce. (Of course, as they are 100% related, or very nearly so, Hamilton's rule [see Question 25] happily allows this.) Multicellularity also required the evolution of stem cells: these are undifferentiated cells that, in the adult, can transform themselves into any kind of cell (usually depending on their physical location in the organism) to replace cells that are dying, thereby extending the life of the organism so as to provide time for reproduction. (They are also crucial during embryological development, of course, as they are the cells that give rise to the different types of cells as these develop—not to mention a possible medical role in regrowing parts of the body that have been destroyed by accident or disease.)

Multicellularity evolved because it offered a range of advantages, including more efficient sharing of nutrients, more effective resistance to cellular predators (most of which simply engulf their prey), the capacity to attach themselves to a substrate in order to remain in one place (especially in watery environments where currents can sweep you away), and the capacity to reach upward to filter-feed or to obtain sunlight for photosynthesis. What makes this possible is that different cell lineages engage in a division of labor, one providing bone for skeletons, another muscles, a third blood, a fourth nerve tissue, and so on. By cooperating, they all benefit on a "the whole is greater than the sum of its parts" basis because they can achieve benefits of a kind or on a scale that the individual cells on their own could not hope to achieve.

However, multicellularity also has costs. The need to regulate the growth and stasis of whole classes of cells is a finely balanced process and can easily be destabilized. When that happens, the result is cancer, in which certain cells run amok and continue to divide when they should have stopped. Cancers appear to be unique to animals; they do not seem to occur in plants. It is not known why this is so. However, the fact that vertebrates (in particular) have many more cell types than plants (making the management of cell growth and death a more complicated task) is likely to be part of the answer.

35. What is symbiosis?

In the course of evolutionary history, different species have occasionally come together to form a close partnership known as a *symbiosis*. Among the best-known examples are lichens, which are not a single plant but a stable symbiosis between a fungus and an alga (or in some cases a cyanobacterium). Like all fungi, lichens require carbon as a food source, and this is provided by the algae or cyanobacteria via photosynthesis.

Some vertebrates can even be symbionts: cleaner wrasses and gobies, and even some shrimps, provide a parasite-removal service for giant oceanic fish like groupers, marine iguanas, and sometimes even corals. The host provides the cleaner with dead skin and ectoparasites (surface parasites on the skin) as food, and the cleaner provides a hygiene service that helps the host keep its skin in good condition.

In some cases, only one of the species benefits from the arrangement, with the effect on the other being neutral or minimal. These cases are sometimes distinguished as *commensals*. Well-known examples include plant galls: the gall is a hollow growth from the tree bark within which lives a particular species of ant, wasp, or even a virus. When laying its eggs, the invasive species often injects a chemical into the plant to stimulate gall formation. Most of these are harmless to the plant. However, in some cases, the symbiont can provide a useful service for the plant by attacking herbivores so as to discourage them from eating the plant's leaves and stems. Other examples include the remora fish (that attaches itself to the skin of sharks and other large fish to use the host as transport, often feeding on scraps of their host's meals), many of the epiphytic plants (such as mistletoe and the many species of orchids that grow on trees and extract nutrients directly from them), leafcutter ants that farm fungi in their nests as a food source—and, of course, the dogs and cats that exploit us (though we did admittedly have a hand in their evolving this lifestyle, and perhaps might be said to gain a benefit from them in return for giving them food, shelter, and walks).

As noted earlier (see Question 24), the genome of most species has a relatively small number of functional genes. Of the two billion codons (genes) in the human genome, only about 20,000 are functional (i.e., code for proteins). A significant proportion of the rest consists of viruses that inserted themselves into our DNA at various stages over the last three billion years. Since they have no direct effect on us either way, they are able to hitchhike their way through history relying on us to reproduce them.

Ruminant mammals (cows, sheep, deer, antelope) offer an example in which symbiont bacteria provide an essential service. These species have all evolved as specialist herbivores, adapted for living on grass and, in some cases, the leaves of bushes. Grass and leaves have unusually tough cell walls, designed to allow them to survive desiccation in dry summers, especially in open habitats where they are exposed to direct sunlight. The lignins that make up the cell wall of grass leaves, which are similar to the lignins that give tree trunks their strength, are practically indigestible for animals. Instead, all the species that eat grasses, from termites to cows, rely on bacteria in the gut to digest the cell walls for them. The host can then access the nutrients by digesting the bacteria or their waste products. The bacteria get a benevolent environment in which to reproduce and multiply, and the herbivore can survive in habitats that would otherwise be impossible to occupy. What could be better?

36. What symbionts are important for us?

Perhaps our most important symbiont is the mitochondrion. Mitochondria live within the cytoplasm, the gel-like substance inside a cell. There are exceptions: red blood cells, for example, have no mitochondria at all. Although mitochondria are involved in a number of cellular processes, their most important function is to provide the cell with energy, which they do in the form of the chemical ATP (adenosine triphosphate) through respiration—something the cell itself cannot do. The process by which it does this is known as the citric acid cycle (or Krebs cycle after its discoverer, the Nobel laureate Hans Krebs).

Although there are alternative suggestions, the consensus view is that mitochondria were prokaryotic bacteria that invaded a eukaryote cell at a very earlier stage and survived by providing the host cell with an independent source of energy.

On the basis of similarities in genome, they seem to be related to *Rickettsia* bacteria, which have to invade the cells of living organisms (normally via an intermediate host such as ticks, fleas, or lice) in order to survive and reproduce.

The second most important class of symbionts we have are the bacteria and other microbes that live in our guts. It has been estimated that the average human gut contains around 30 trillion bacteria from dozens of different species—almost the same number of cells as in the human body itself (which has around 39 trillion cells). These gut microbes include archaebacteria, bacteria, fungi, protozoans, and viruses, although around three-quarters are made up of just one class of bacteria. The combined DNA of these gut microbes is 100 times the size of our own DNA. In fact, our guts are a battlefield of microorganisms as the different species compete with each other for access to what is a veritable bacterial nirvana.

Most of the bacteria in our guts are absolutely essential for us. Without them, we would be unable to acquire the nutrients we need to maintain life. Like the gut microbes that allow ruminants to digest leaves (see Question 35), they help process nutrients into a form that we can absorb through the gut wall. For a species like humans that has an energetically very expensive super-large brain, this is crucial. They also help train the immune system to distinguish between friendly bacteria and viruses and the pathogenic ones that daily invade our bodies. Without them we would be cognitively deficient, emotionally unstable, and at the mercy of the many diseases that constantly assail us.

Our initial colonies of gut microbiota are acquired during the birth process through skin-to-skin contact with the mother's vagina as we are being born. Of course, we add to this complement throughout life as we eat and have close physical contact with other individuals. It is estimated that we exchange 80 million bacteria during a 10-second kiss. Nonetheless, the complement first established at birth has a profound effect on our early development. Babies born by caesarean section have a deficient complement of gut microbiota, and this results in a cognitive development that is noticeably delayed or even anomalous.

Perhaps the most interesting of the microorganisms that live with us is the tuberculosis (TB) bacillus, which is normally viewed with trepidation and fear as a major killer. In fact, the TB bacillus may be more of a friend than we think. It provides an important source of niacin (vitamin B_3) as a waste product of its metabolism. Niacin is essential for brain growth and maintenance through its effect on the processing of fats, carbohydrates, and proteins. Without it, we would rapidly suffer brain degeneration in the form of diseases like pellagra. Even mild deficiencies are associated with nausea, dry skin, tiredness, and headaches.

Niacin is not widely available in plants (but it is present in many of the spices we eat!). In order to ingest the quantities we need to support our super-large brains, we have depended on eating meat, which is unusually high in niacin. Meat is hard to acquire in nature, however, and supplies can be erratic. Fortunately, the TB bacillus produces copious quantities as a waste product. Niacin is poisonous for the bacillus (overdosing the bacillus on it was one of the earliest successful treatments for TB), so it excretes it into gut and other tissues where it lives. We are then able to exploit this and turn it to good advantage. As with all things in biology, of course, things can go wrong, and when they do the result is that the bacillus runs riot, resulting in the sometimes fatal symptoms of TB.

37. When is an individual a cell, an individual, or a colony?

This question is actually the biggest philosophical conundrum in the whole of biology. Darwin's theory of evolution by natural selection focused on the individual as the unit of selection. The individual makes sense to us, because we experience ourselves as an individual and we see many individuals walking around or growing in the environments where we live. But it turns out that not all organisms fall so neatly into the simple individual-made-up-of-genes mold. In some cases, things can be a great deal more complicated.

Slime molds can switch from living as individual cells (when food is plentiful) to living as a multicellular organism with specialized cells (when food is in short supply). The multicellular phase is typically initiated by a chemical signal given out by some of the cells. At this point, the individual cells fuse into a single large bag of cytoplasm with many gene nuclei (known as a plasmodium) that flows across or through a substrate in search of better living conditions. The plasmodium can achieve speeds as fast as 8 meters (27 feet) in an hour. Although slime molds in this phase vary greatly in size, some can be as large as 1 meter (39 in) long.

When food conditions deteriorate completely, slime molds in the multicellular phase form specialized stalks that produce fruiting bodies that release large numbers of spores. The spores are carried off in the wind or on the coats of passing animals in the hopes of finding somewhere better to settle. The stalks then die away and their cells thus do not themselves reproduce. In doing this, the slime mold behaves like a true multicellular organism with specialized non-reproductive cells. Meanwhile, when the spores find suitable conditions, they "germinate" into new single-celled amoebae and the cycle starts again. If a slime mold mass is physically separated, the two halves can find their way back together and reunite. Is this an individual or a cooperating community? It is not obvious, and our everyday experience does not give us any ready-made criteria for deciding.

The plant world offers us many other equally confusing examples. Many plants clone themselves by sending out roots that sprout a new plant—or what looks like a new plant when we see it from above ground. But in fact, underground it is all one enormous plant. Many familiar garden plants like the strawberry and many grasses reproduce in this way. Some trees even behave like this. The Huon pine of Tasmania, which is genetically male, reproduces by sending out subterranean roots that sprout new stems. The King's Holly, another Tasmanian tree, consists of just one clone in the wild with 600 individual stems covering an area 1.2 km (3/4 of a mile) in length. Unusually, it has three sets of chromosomes instead of the normal two and so is sterile. It reproduces vegetatively: whenever a branch breaks off, it has the capacity to send out roots and establish itself as a new individual. But genetically, it is identical to its parent. In the Wasatch Mountains in Utah, there is a population of 47,000 quaking aspen trees that turned out to be a single clone; this one individual covers an area of 43 ha (106 acres), and is estimated to be around 80,000 years old.

Bacteria and other prokaryotes provide another challenge. They look like individuals when we see them at any given moment. But as they reproduce asexually by dividing into two separate cells (individuals), the concept of individuality once more becomes strained: all the descendants of one ancestral cell are more or less identical to it (barring the occasional mutation) and so form what you might think of as a temporal clone—a bit like a slime mold in its solitary phase. The same is true of the many invertebrates that reproduce asexually. Huon pines and quaking aspen are simply at the other extreme: they just like to keep all their offspring attached.

These examples challenge our everyday concepts of individuality. But they don't particularly surprise biologists. Although Darwin thought in terms of individuals as we commonly understand them, it is genetic lineages that actually lie at the root of his theory as was later appreciated by Hamilton (see Question 25) and the modern synthesizers. I belong to a genetic lineage that includes all my ancestors and extended family, which will continue down through our collective descendants. As mammals, we choose to mix our genes with those of other individuals (though, like all animals, we typically prefer individuals who are as similar to us as possible). But that is really just an accident of the specializations that evolved as these genetic lineages evolved more and more complex ways of coping with environmental challenges.

38. So am I an individual or a colony?

This is really a very good question. The fact that most of us are a curious mosaic of different kinds of cells (see Questions 34 and 36) raises deep philosophical questions about who we actually are. Are we a single individual (as our conscious mind seems to tell us), or are we more like colonies of coral, a whole host of individual cells bound together by some kind of biological cement?

In fact, most of your body is constantly replenishing itself as cells die off, or even self-destruct, at the end of a carefully managed life. The cells in the surface layer of your skin (the epidermis) are constantly being rubbed off and are replaced by new cells growing from underneath, so that your surface skin is replaced on average every 27 days. (This is not true, incidentally, of the internal layer of the skin, the dermis. That's why tattoos created by injecting ink into the dermis remain for longer than you usually would like them to.) In fact, by the time you have finished reading this sentence, 50 million cells in your body will have died and been replaced. With a few exceptions, most of the cells in your body are replaced every 7–15 years. In short, you are not actually the same physical person you were 10 years ago, even though you may *feel* that you are.

The fact is that you are a complex mosaic of cells: some are so specialized their only future is to die with you, some have the potential to be immortal (providing you breed), and about half (the bacteria in your gut) (see Question 36) aren't even related to you. This raises an interesting evolutionary conundrum that lies at the heart of every multicellular organism: why should all these cells cooperate with each other rather than do their own thing? If DNA behaves selfishly, as the selfish gene hypothesis (see Question 26) suggests it should, why don't these cells just go mad and replicate on their own as fast as they can?

One obvious answer is that this a classic case of evolutionary cooperation—in some cases between very closely related cells, and in other cases between completely different species. The evolution of multicellularity is premised on the fact that by cooperating, their genetic lineages stand a better chance of being represented in the next generation. Your own cells are behaving a little bit like the worker bees in a beehive (see Question 72). They are close to 100% related to each other (the odd mutation here or there notwithstanding), so it should benefit them to assist in the reproduction of their lucky gamete sister cells. But gut microbiota aren't related to you at all, and they should be much more interested in exploiting you for their own ends. That can happen.

One such case was discovered by Alfred Russel Wallace himself during his collecting expeditions in the forests of Borneo. The *Ophiocordyceps* fungus attacks carpenter ants that live in the trees. Once infected by the fungus, the ant is driven to descend to the ground where the humidity and temperature are suitable for fungal growth. There, the ant finds the underside of a leaf to grip onto with its jaws and goes into a kind of suspended animation until it dies 7–10 days later. Meanwhile, the fungus grows until it sends out fruiting bodies that burst out of the ant's head, and eventually release spores.

Another example is the *Myrmeconema* nematode (a tiny worm). When it infects a canopy ant, it causes the ant's abdomen to change color so that it resembles a ripe fruit. The "fruit" is then eaten by a bird, whose droppings are later collected by more ants and carried off (along with the nematodes they contain) to their nests, allowing the nematode to find new hosts. How weird is that? A more familiar case is the rabies virus that causes its host (including humans, on occasions) to become more aggressive so that it will bite a random victim, allowing the virus to be transmitted to a new host via the old host's saliva.

Fortunately, not many of our colony of microbes, never mind our own cells, are so devious. Their various self-interests are mainly aligned, and that helps to keep the lid on the system. In part, this is a consequence of a long evolutionary history in which the various components adjust to each other's requirements as symbionts (see Question 36).

39. Why don't we live forever?

All organisms, from bacteria to humans, grow old and die. So an obvious question is: shouldn't natural selection have found a way of allowing us to live forever? After all, an individual that lives longer will inevitably produce more offspring than one that lives less long. Some tree species live to be many thousands of years old (see Question 37), so why can't we?

In fact, we humans are often so desperate to live forever that we are prepared to spend vast sums of money on medical treatments, diets, and exercise regimes allegedly designed to achieve that end—in some cases even having our bodies deep frozen in the vain hope that, at some future time, medical science will have figured out how to bring us back to life. There are three fairly obvious reasons why we might expect to be disappointed, notwithstanding the power of natural selection to bring about any outcome that is advantageous.

One is that there is a trade-off between investment in reproduction and investment in somatic survival: an organism can choose to invest everything in one massive throw of the reproductive dice at the expense of its own survival. This is what species like the Pacific salmon, the male marsupial mouse, the marine bristle worm, and the West African raffia palm actually do. They throw everything into one single reproductive effort, and then die exhausted. The problem is that reproduction is very expensive in terms of energy, and no organism has infinite energy. Sooner or later, the energy costs of reproduction will catch up with you. The sex hormone testosterone imposes considerable stress on the male physiological system, and the resulting wear and tear significantly reduces male life expectancy. An analysis of longevity in the eunuchs of the Korean royal court over the last 500 years revealed that, on average, they lived 15–20 years longer than uncastrated men of similar social and economic status. Similarly, childbirth and rearing impose significant costs on women, and at least one demographic study of a historical European population found that a woman's longevity was inversely correlated with the number of children she gave birth to.

A second reason is that bodies just wear out. As we saw earlier (see Question 38), body parts are constantly being replaced. Some parts of the body are less easily replaced than others, and when these wear out there isn't much that can be done about it. More importantly, mutations are likely to accumulate with each cell replication. In addition, genes with adverse consequences for the organism that express themselves late in life will be less exposed to selection than those that express themselves earlier in life: the latter will kill their bearer before they have had time to reproduce, whereas the former will have reproduced before natural selection has had time to act. These will catch up with you in the end, and there is nothing you can do about it.

A third reason is simple accidents of evolution. Some organisms have a material construction that is especially tough and this allows them to live much longer because they are less likely to get damaged. Trees are among the longest-lived life forms on earth. Some California redwoods are estimated to be as old as 2,000 years. Some individual stems of the Tasmanian Huon pine (see Question 37) are also 2,000 years old, while the whole clone may be as much as 11,000 years old. But even these examples pale into insignificance by comparison with its neighbor, the King's Holly, whose sole surviving clone is around 43,000 years old.

One of the reasons that trees can live so long is that the lignins (and especially the recently discovered nanocrystalline cellulose, which is eight times stronger than steel) that make up the fibers of their trunks are particularly tough, and able to resist both accidental breakage and attack by herbivores. Most organisms, however, don't have this advantage, because a body made out of these materials has two important disadvantages. First, it makes you *very* heavy, which makes movement increasingly expensive. (That is one reason why trees don't walk about, the *Lord of the Rings* stories notwithstanding.) Second, tree trunks are relatively inflexible, which is why they don't bend too much in the face of even the strongest winds. If you want to be able to twist and turn while clambering about in the trees or escaping from your predators, you don't want to be a tree.

It's all about trade-offs between different options, any of which may be reasonable alternative solutions to the problems of survival and reproduction. The complexity of these biological decisions makes it all but impossible to find a perfect solution that beats all others. In short, there are usually many different ways to maximize fitness. The best that natural selection can ever achieve is the best of a bad job.

40. Even if evolution is true, why should a theory of evolution be relevant for us today?

Perhaps the most pressing practical reason for understanding the processes of evolution is disease. The single most important problem that we face is the rate at which new diseases appear for which we have no medical solutions. This applies as much to the diseases that afflict us (from Ebola to bird flu and the Zika virus) as to the diseases that afflict our crops and farm animals—the 7,000 species of rust fungi that blight our domestic crops, or the cyst nematodes that cause $300 million worth of damage every year to European potato crops.

Understanding evolutionary processes can allow us to eradicate diseases more effectively, as was done with smallpox in the 1970s. But more often, our problem is one of just keeping up with the speed at which viruses and bacteria change under the impact of natural selection. New diseases and new versions of old ones appear with monotonous regularity, year on year. They threaten both our health and our ability to feed ourselves. Our treatments invariably work only for a while, and then spawn new strains that are resistant to everything we throw at them. Examples have included myxomatosis, a disease that was used to control rabbits and hares in the 1950s, MRSA and the many other antibiotic-resistant superbugs that have bedeviled our hospitals in the last couple of decades, DDT-resistant and chloroquine-resistant malaria, and the list goes on. Without a theory of evolution, we will always be caught out by new diseases and will have much more trouble trying to deal with them.

The story of malaria resistance is instructive. DDT, an organochlorine compound, was first synthesized in 1874 and its insecticidal properties were discovered 65 years later by the Swiss scientist Paul Hermann Müller. It was used extensively by the military (in particular) during the Second World War to control typhus and other diseases among both troops and civilian populations during the chaos of the war and its immediate aftermath. So successful was it that Müller was awarded a Nobel Prize in 1948. It was then widely used in the tropics during the 1950s to control malaria by killing off the mosquitoes that were the carriers of the malaria plasmodium. However, by the 1960s, it had become apparent that, aside from the fact that DDT was poisoning the rest of the environment (see Question 48), the malaria plasmodium was becoming immune to its effects. Much the same story emerged in the 1980s with chloroquine, the main antimalarial drug used both as a prophylactic and as a treatment following its synthesis in the 1940s. The plasmodium has developed a single gene mutation in the PfCRT gene that allows the cell to excrete rather than digest the chloroquine that is meant to poison it. Chloroquine resistance spread rapidly throughout most of the New and Old World tropics, mainly of course thanks to mass travel made possible by the airplane.

Another reason we might be interested in the implications of evolution is in understanding our own behavior. We are here; to be here, we must have evolved, and how we evolved must have been subject to the processes of evolution. As with all things biological, some aspects of our behavior will be relatively fixed, and some will be very flexible and open to learning or manipulation. Understanding which aspects of our behavior and psychology are flexible—and which not—might be important for understanding whether we can change human behavior for the better, and how best to do it. And if we cannot change it, how might we manage it so as to minimize its adverse effects on society.

All these practical uses notwithstanding, the ultimate reason for having a theory of evolution is the sheer joy of understanding the natural world, how it works and how it came to be. Curiosity gives pleasure—and it also spawns new hypotheses and theories as we discover new facts about the world and struggle to explain them. That can lead to unexpected discoveries that contain the seeds of new solutions to some of the problems of survival and reproduction that dog us. Without a theory of evolution, we wouldn't even think of asking the kinds of why, when, and how questions that give rise to understanding (see Question 10).

5

EVOLUTION OF SPECIES

41. What is a species?

Having some way of classifying the natural world is important. That is the starting point of all science. For biology, it is also important for undertaking the kinds of very sophisticated statistical analyses that involve comparisons across different species and genera—the method that, in less sophisticated form, Darwin himself used to great effect. When we do such analyses, we need to be sure that we are comparing like with like in terms of species relatedness. It is important that our taxonomies (the treelike structures that describe how species, genera, families, orders, and other taxonomic groupings are related to each other) map as closely as possible onto evolutionary history. There are also practical considerations: if two populations of organisms are under threat but we can realistically save only one of them, should we be less concerned if they belong to the same species than if they belong to different species?

The first formal taxonomy was developed by the Swedish anatomist Carolus Linnaeus during the late eighteenth century (see Question 5). His taxonomy was based on anatomical similarity (on the reasonable assumption that species that looked similar were likely to be related to each other). He used a commonsense definition of *species*—a collection of individuals that looked pretty much the same. Cows and horses are quite obviously different kinds of animals, but sheep and goats look more similar to each other and so are likely to be more closely related. Linnaeus established a *type specimen* for each species (he used himself as the type specimen for the human species, *Homo sapiens!*) on the grounds that there is an ideal form for each species that every member of the species does its best to match up to (a concept originally derived from the great Greek philosopher Plato, and now referred to as *Platonic types*). In effect, every member of the species has the same basic template, and the differences between individuals of the same species are just the imperfections of real life.

While this Platonic definition of species has worked fine over the last two centuries, it began to unravel during the mid-twentieth century as a result of two key facts. One was increasing evidence that animals from different species could, and did, mate with each other, and sometimes produced fertile hybrids when they did so. The carrion and hooded crows, for example, seem to be good species by anyone's definition, but they hybridize readily where their ranges overlap. Modern humans demonstrably interbred with Neanderthals in Europe before the Neanderthals went extinct, so that most Europeans have a small proportion of Neanderthal genes (see Question 65). Platonic types are evidently not quite as unique as we had assumed. Indeed, some species have turned out to be mosaics of two or more species (e.g., lichens, of which there may be as many as 20,000 species) (see Question 35).

The second problem was that this definition jarred rather badly with the evolutionary modern synthesis and the new population genetics. Evolutionary theory views a population as a collection of related individuals whose traits (and, of course, genes) all differ very slightly as a result of mutations acquired over the generations back to their common ancestor. Thus, rather than the small differences between individuals being due to errors of development of no particular interest, they are actually the whole point. They are what gives rise to new species in the future.

During the 1950s, the emerging consensus was that a species should be defined as a set of animals that were capable of interbreeding to produce viable (i.e., fertile) offspring. Horses and donkeys can interbreed, but the progeny (a mule) is infertile, so this makes the parents two good species. This became known as the *biological species definition*. It recognized that in nature, the genetic differences between all living organisms form more of a lumpy continuum rather than being completely discrete. The fact that they do not always produce fertile hybrids when they occasionally interbreed in the wild is usually due to the simple mechanical fact that their chromosomes do not match up well enough when the sperm of one species fertilizes the egg of another. The garlic plant has just 10 chromosomes, for example, whereas wheat has 42 and the adder's tongue fern has over 1,000. Among animals, the muntjac deer has only 7, whereas humans have 46, chimpanzees 48, dogs 78, and the kingfisher 132. Although this is not always an impediment, a fertilized egg typically won't develop properly if the chromosomes from the two parents cannot pair up with each other. When this happens, the embryo will soon die.

42. Why are species sometimes difficult to define?

Although the biological definition of a species (see Question 41) works well enough, it isn't perfect. There are many exceptions, especially among plants and microscopic life forms. The conventional definition of a species assumes that every individual gets its genes only from its two parents, but this isn't always true as there can be horizontal transfer of genetic material between genetic lineages through bacteriophages (a type of virus) and plasmids (a small DNA molecule, sometimes also known as a *replicon*, that exists within a cell and replicates separately from the chromosomal DNA). These can invade a new host's cells, bringing foreign DNA with them. This seems to be especially common in prokaryotes. Indeed, this may be the primary mechanism for the acquisition of antibiotic resistance in bacteria (see Question 40). But it is also known to occur in crustaceans (the crabs and their allies) and the echinoderms (sea urchins and sea cucumbers).

There can be awkward cases even among large-bodied vertebrates. In the case of fossil species across long periods of geological time, it can be difficult to decide when a species has changed so much that it probably should be counted as a new species. "Ring species" are another source of problems. One of the best-known examples of this are the seagulls of northern Europe. In western Europe, these start off as our familiar lesser black-backed gull. But these shade eastward into the Siberian gull, then into the vega gull of eastern Siberia, across the Bering Strait into the Smithsonian gull and finally into the herring gull, which lives happily as a distinct and separate species alongside the black-backed gull in western Europe. Yet each adjacent pair of species going eastward is quite capable of interbreeding. The ring is thought to have been created by the fact that winds tend to blow west to east around the North Pole, so that birds are naturally pushed eastward. Since they interbreed more easily with species to their east, there is eastward-flowing gene transfer such that, by the time it gets back to Europe via America, there is sufficient accumulated genetic divergence for it to be impossible for the species at the two "ends" of the ring to interbreed.

The truth is that we are dealing with a natural world that is continuous: in reality, everything *is* related to everything else by descent from a common ancestor, even if the chain of relatedness is sometimes several billion years long. Our problem is that we are still trying to fit the continuum of this natural variability into square Platonic typological boxes. That is mainly a consequence of human psychology: we just find it much easier to think typologically. Continua are much more difficult to deal with. But the real world of animals and plants consists of individuals that are simply more or less closely related to each other. Defining individual populations as species or genera, or other higher taxonomic groupings, is thus as much a matter of convenience as biological rigor. Biologists need to be able to refer to the organisms they study, and they need to know whether or not the ones they study are (roughly) the same as those studied by someone else, and, if not, how closely related they are. A species concept is useful for this, but we shouldn't get too carried away with it. One could even argue that it is the least interesting thing in biology (though I wouldn't say that to a taxonomist if you want to live a long and happy life).

Despite all these drawbacks, the biological definition of species works for us most of the time. Bacteria do not mate with humans, any more than birds mate with mammals. Once we have sufficient distance between species, the biological definition tends to keep the lineages pretty distinct. One problem, though, is that we have phases of enthusiasm for creating new species. These are usually interspersed with phases when we radically reduce the number of species by recombining the more closely related ones. People seem to divide naturally into enthusiastic splitters and enthusiastic lumpers.

We tend to become more enthusiastic about splitting the closer we get to humans. What would count as different genera in primates would barely count as different species in beetles. Primates provide a particularly glaring example. In the 1980s, around 120 different species were recognized; by the 2010s, however, some people were claiming there were as many as 350 species—mainly because what had previously been a single species has been split into a dozen or more species because of slight genetic or pelage differences between populations. There may be political advantages to increasing the number of species for practical conservation purposes (and personal benefits for being cited as the authority), but it is questionable whether there are always convincing scientific grounds for doing so. That does not, however, stop us from being interested in Darwin's original question: how do new species (however we define them) arise?

43. How do new species arise?

In the century and a half since Darwin published *The Origin of Species*, the processes of speciation have remained of central interest to biologists. Darwin himself had provided much of the answer. Populations that become geographically isolated acquire specialized adaptations to their local environments; eventually, they evolve appearances that are sufficiently different for us to want to class them as different species because they *look* so different.

Since then, studies of speciation have incorporated a better understanding of genetics and the mechanisms of inheritance, and these have refined our understanding of the processes involved in several important ways. One of these is that the level of gene flow between two populations in the process of becoming separate species should be minimal. If there is too much gene flow as a result of matings between individuals of the two populations, the gene pools will be unable to differentiate because any genes that are lost in one population will be replaced by influxes from the other, and any new genes that appear in one will quickly find their way into the other. This is often a consequence of members of one population moving between the populations in search of mates.

An ancillary requirement is geographical separation (technically known as *allopatric speciation*). It is much easier for a species to split into two genuine daughter species if their ranges have become separated because individuals cannot move between them. This might happen if a major river alters its course, a rift valley opens up down the middle of the species' range, or volcanic islands drift apart (the various species of Darwin's finches and the giant tortoises on the different Galapagos Islands are an example of this). When this happens, the new species might arise through local adaptations or simply drift apart genetically as they accumulate new mutations (see Question 29).

A particularly important element in allopatric speciation is the *founder effect*. Most new species arise through a small subset of the parent population moving into a new environment on the edge of the species' geographical range. There, it may undergo rapid evolution, causing it to diverge very quickly from the parent population. This can happen if the new habitat places it under very intense, novel selection pressures. But it can equally happen simply because the migrating subset will not be fully representative of the genetic variation in the parent species as a whole. Its peculiar genetic quirks are then likely to become magnified within the new species as a result.

Speciation can also happen if there is a mass extinction of the parent species' populations and only one or two small peripheral populations survive. Modern humans seem to be an example of this: we all descend from a population of about 5,000 breeding females who lived somewhere in northeastern Africa around 150,000 years ago. This is known as the *Mitochondrial Eve hypothesis*, so named because the evidence for it came from examining the similarities and differences in the mitochondrial DNA of women sampled from all around the world, mitochondria being inherited only down the female line (see Question 36).

Speciation can sometimes occur even if there is only partial geographical separation. This is known as *parapatric speciation*. However, it usually requires sufficient differentiation within the gene pool of a population for the hybrids formed by matings between members of the two genetic subtypes to be less fit than either of the parents. When this is the case, there will be reduced gene flow between the two halves of the species' gene pool, and eventually they will become two separate species. This normally requires a habitat gradient of some kind that allows local adaptation to different habitat types that is big enough for the populations at the two ends of the distribution not to meet up very often. Ring species (see Question 42) might be an example of this.

The third mode of speciation is known as *sympatric specia-tion* because the two incipient species actually live in the same habitat. This is quite common in parasitic species, where individuals within a population start to specialize on different hosts. It still requires reproductive isolation to occur. In North American hawthorn flies, some populations started to exploit apples after these were introduced from Europe during the nineteenth century; eventually, they became apple specialists unable to interbreed with the hawthorn specialists. The North American snow goose may also be an example: it has two color morphs (blue and white) that are controlled by a single gene (blue is dominant, white the homozygous recessive) (see Question 22), but both breed in the same areas of the Canadian Arctic. Although they do sometimes interbreed (and produce fertile offspring), there is a tendency for color segregation to occur (blues prefer to pair up with blues, and whites with whites). This entirely accidental segregation (presumably the result of cues of kin recognition) (see Question 27) will eventually result in two separate species if interbreeding rates are low enough for long enough.

One form of sympatric speciation that may have been quite common, at least in plants, is *polyploidy*. Polyploidy involves the accidental duplication of chromosomes during meiosis, such that the offspring ends up with multiple copies of all or some of its chromosomes. It is thought to be the main genetic mechanism whereby species have evolved large numbers of chromosomes (see Question 41). Since individuals with different numbers of chromosomes do not usually produce offspring when they mate, there is what amounts to instantaneous reproductive separation between subpopulations—at least, providing there are enough individuals with the polyploidy to mate with each other. The presence of extra chromosomes usually results in a dramatic change in the offspring's appearance (its *phenotype*).

Polyploidy of a limited kind is most familiar in humans in the form of Down syndrome, which results from an accidental duplication of chromosome 21 so that the individual ends up with three copies instead of two (known as *trisomy 21*). This usually occurs when both copies of the chromosome accidentally segregate together during cell division, so that one daughter cell has two copies and the other none.

44. Why do genetics and anatomy sometimes disagree about evolutionary history?

In general, the phylogenies (or evolutionary family trees) generated by the evidence from molecular genetics are comfortingly similar to the traditional ones based on anatomy. In general, they agree completely about the major divisions (amphibians versus reptiles, birds versus mammals, primates versus carnivores or ungulates), and about most of the major groupings within different orders (within the order primates, between prosimians and anthropoids, or between apes and Old World monkeys). Where they more often disagree is in the exact relationships between species within closely related genera. Taxonomists have long recognized five species of African baboons on anatomical grounds (they do *look* different), but genetically they are a confused mess, probably reflecting gene flow across geographical boundaries. Some people would argue that they are best considered a single species, despite their quite recognizable physical differences.

There have, however, been some spectacular disagreements that have forced us to change our understanding of evolutionary history. Perhaps the best known of these is the position of modern humans within the hominids (the family of large ape-like species, excluding the gibbons or lesser apes). On the basis of anatomy, taxonomists always viewed humans as the outliers in this group because of their bipedal stance, much larger brain, and other traits such as hairlessness, speech, and tool-making abilities. Chimpanzees and gorillas were obviously quite closely related (forming an African great ape family), and the Asian orangutan was clearly a closely related sister species to this group. However, in the 1970s, the genetic evidence turned all this on its head. In fact, it was the orangutan that was the distant cousin within the family. Chimpanzees, gorillas, and humans form a single African great ape grouping, within which the gorilla as a slightly less closely related sister species for the two chimpanzee species and humans. Despite all our anatomical and psychological differences, we *are*, in Jared Diamond's memorable phrase, merely the "third chimpanzee."

The confusion had been a consequence of not recognizing that strong selection pressures can result is rapid phenotypic change without major genetic change. This occurs when just a few key genes that determine aspects of a species' appearance change, but the rest of the genome remains unaffected. Many of the seemingly dramatic differences between humans and great apes turn on changes in just a handful of genes, not in large numbers of genes. Appearances can be deceptive.

Primate taxonomists were later given a second shock when the genetic evidence revealed that the gorilla, which had always been viewed as a single species, turned out to be two separate species that had diverged around 260,000 years ago (i.e., before our own species separated from the other archaic human populations) (see Questions 30 and 61). This seems to have been a classic case of allopatric speciation. When the Ice Age caused the forests of central and western Africa to retreat into a small number of isolated forest patches, the various gorilla populations got trapped in different places. During the time that these forest blocks remained separated, their populations followed slightly different genetic trajectories and diverged through genetic drift or local selection—but with only minor effects on their physical appearance.

This is also the likely explanation for the profusion of guenon monkey species that now often coexist in the same central African forests (and often forage together in the same multispecies groups). These monkeys seem to have evolved in isolated forest blocks after these became isolated during the last Ice Age. Eventually, as the Ice Age drew to its close and the forest blocks joined up again, the new species encountered each other but were now able to coexist because they differed sufficiently not to compete with each other.

Microspecies, cases where a species has a range of geographical varieties that look different but do not differ much genetically, are in fact quite common. Examples include the dandelion (a single species with some 400 microspecies) and the blackberry (with 200). Animal examples are offered by *Heliconius* butterflies and *Hypsiboas* treefrogs, and perhaps the five (or six) species of the baboon that we met earlier. Many of these may well be populations in the process of speciation (they will eventually form genuine new species if left to their own devices). And this perhaps reminds us of an important lesson: what we see now in nature is but a snapshot of a dynamic process, with different lineages caught at different stages of their evolutionary histories.

45. Why do species go extinct?

Species, and individual populations, go extinct for the rather obvious reason that they do not produce enough babies to compensate for the rate at which individuals are dying. The underlying question, then, is: why are the adults dying so fast?

The main reason that mortality rates suddenly become high is invariably climate change—or its consequence, changes in local vegetation conditions or prey availability. Most species become adapted to particular environmental conditions, and if these can change faster than populations can track them, extinction will beckon. Gelada baboons, for example, were a widespread genus found all over the grassland habitats of sub-Saharan Africa, as well as southern Europe and Asia as far east as India, until perhaps half a million years ago. Then they disappeared quite suddenly, leaving just one very localized relict species on the high-altitude plateau of northern and central Ethiopia. Geladas are unusual among primates in being grazers, and are adapted to the soft montane-type grasses that were formerly widespread at much lower altitudes. This seems to have been a consequence of the climate change that kicked in around two million years ago with a dramatic drying of the climate, resulting in the shift of this grassland belt up to cooler high altitudes and its replacement at lower altitudes by the more heavily silicated grasses now characteristic of Africa's savannahs. Geladas cannot cope with these tough, silicated grasses, and in the lower-lying habitats could not move fast enough to track the changing geographical distribution of their preferred habitat. They just died out.

This example illustrates one of the predisposing factors that makes extinction more likely: namely, dietary specialization. As one of the dietetically most specialized primates, geladas were at particular risk. In contrast, the omnivorous and ecologically much more flexible common baboons expanded their geographical range considerably as the gelada populations were dying out, replacing the geladas throughout most of sub-Saharan Africa. One modeling exercise of the effects of future climate warming on primate populations showed that folivorous species (those that, like the spectacularly beautiful African colobus monkeys, specialize on leaves and eat less fruit) will struggle to survive, whereas the more omnivorous fruit eaters will, broadly speaking, be fine. The problem folivores have is that eating leaves is a time-costly business (you have to set aside extra time for fermentation and digestion in the way cows and sheep do), and time is the one thing that most species do not have. Indeed, it is time pressures that most often hamper animals' ability to survive in poor habitats, not the scarcity of food as such (see Question 84).

Another important risk factor is large body size. Becoming larger is an effective anti-predator strategy since it reduces the number of predator species that can eat you. However, growing a larger body requires more time (mainly because cells divide at a constant rate, and a larger body composed of more cells simply requires more cell divisions). This means that development is slower, and the interval between successive births is longer, which in turn means having fewer babies in a lifetime. Large animals are therefore slower to respond when an environmental disaster causes many deaths, and as a result are more susceptible to local extinction before the population has time to recover.

Since lineages tend to increase in size over time in an escalating arms race with predators (see also Question 51), they tend to be driven into an evolutionary cul-de-sac from which it is difficult to escape. Dwarfism (becoming smaller) does happen (an example is the one-meter-tall [three foot] dwarf hominin known as the hobbit that was living on the predator-free Indonesian island of Flores until as recently as 20,000 years ago), but it is relatively much rarer.

Small population size is another risk factor. Small populations have less resilience because they don't have the spare capacity to absorb sudden high levels of mortality (for example, due to floods, volcanic eruptions, or disease). If these small populations are also isolated from each other (for example, on the tops of individual mountains), the risk of extinction is much greater because neighboring populations cannot easily cross the divide to replace each other. This was probably responsible for hastening the extinction of the gibbon populations that lived in central China until the Little Ice Age of the early eighteenth century.

Some locations may also be more risky than others. Areas that are prone to wildfires (initiated by lightning strikes) are risky for less mobile animals that cannot flee the advancing wall of flame. Large rivers that change their course unpredictably (as many do) may also be a problem because they can divide a small but viable population into two smaller (and hence no longer viable) ones. Areas that are prone to hurricanes (the western Atlantic) or typhoons (the western Pacific) are also risky habitats: southeast China is particularly at risk from typhoons, and as a result has an unusually high proportion of species whose conservation status is considered to be "threatened" or worse (i.e., most at risk of extinction).

Last but of course not least, competition with other species can be important. A general principle in ecology is that species cannot coexist if they compete for the same resources: one will drive the other to extinction by eating it out of house and home. Since it was introduced into the British Isles a century ago, the larger, more aggressive, and more adaptable American gray squirrel has eliminated the native European red squirrel from everywhere except a few pockets in northern England and western Scotland. In the 1970s, the American signal crayfish was introduced into Britain by the UK government to create a new freshwater farming industry. Perhaps inevitably, it escaped from the fish farms and spread rapidly throughout Britain's river and canal systems, outcompeting the smaller native white-clawed crayfish as it did so and driving most of their populations to extinction. They also burrow up to two meters into river banks, causing banks to collapse.

However, by far the most serious risk factor in this respect has been our own species (see Question 48). By actively destroying habitats and overhunting species, we are currently responsible for the latest of the several mass extinctions that have occurred in the history of life on earth.

46. How many mass extinction events have there been?

Individual species die out all the time. They become unable to track changes in their environment; or a new predator or disease appears that they cannot cope with. This regular trickle of extinctions gives us the *background rate of extinction* against which to judge whether something more serious is happening. This rate is estimated to be about two to five taxonomic families (clusters of related genera) of marine animals every million years. Mass extinction events occur when the rate at which individual families go extinct rises dramatically above this background rate.

There have been five occasions in the course of the three billion years that complex life has existed on earth in which extinction rates have significantly exceeded the background rate. There may have been others that occurred before the evolution of multicellular life, but these would be difficult to detect; only one, the Great Oxygenation Event (see Question 32), is known for sure. The five known ones have all occurred within the last half-billion years, and took place at about 450, 370, 252, 201, and 66 million years ago. Each marks the boundary between two major geological epochs, and so each is named by the epochs it separates. We appear to be in the grip of a sixth extinction event at the moment.

The first of these, the Ordovician-Silurian event around 450 million years ago, is really a cluster of extinction events, and mainly affected marine communities. Around a third of all brachiopod (a once widespread family of shellfish) and bryozoan families disappeared, as well as many trilobites (one of the earliest groups of arthropods, now long extinct), conodonts (an extinct group of very small-toothed eel-like creatures), and corals. Around 50–60% of all marine genera, and around 85% of all species, disappeared. It is thought to have been due to the dramatic cooling that occurred at the end of the Ordovician era that resulted in a major glaciation (even the Sahara was glaciated at this time!). It resulted in a dramatic fall in sea levels that exposed all the continental shelves, resulting in a buildup of toxicity in the remaining marine environments. Possible causes of the cooling include a gamma ray burst from a hypernova (the gravitational collapse of a star into a black hole) that occurred 6,000 light years away in the Milky Way at about this time. A 10-second gamma ray burst would have stripped the earth's atmosphere of half its ozone layer and exposed surface life to extreme ultraviolet radiation. Another possibility is a burst in volcanic activity around this time that would have cooled the earth by filling the atmosphere with dust, preventing sunlight from reaching the earth.

The Late Devonian event (370 million years ago) affected only marine life and involved two major extinction pulses, the first of which on its own resulted in the extinction of 19% of all families and 50% of all genera. The Devonian was a period of very extensive coral reef building, and almost all of these died out. The causes are unclear, but sea level changes (perhaps due to glaciation again) and declining oxygen levels in the oceans seem to be implicated. A comet impact has been suggested as the trigger, with the Slijan Ring in Sweden (52 kilometers [32 miles] in diameter) being a possible contender for the location of the impact.

The End Permian event (252 million years ago), also known as the "Great Dying," was the earth's most climactic extinction event. An estimated 96% of all marine life died out, as well as 70% of all terrestrial vertebrate species. It is the only known mass extinction that seriously affected insects. Species with calcium carbonate shells were particularly badly affected. The biodiversity of terrestrial habitats took around 10 million years to recover. The pattern of extinctions is strongly suggestive of hypoxia (lack of oxygen). There is some evidence (shocked quartzes, meteorite fragments, and fullerenes [minute carbon spheres] containing noble gases that only occur elsewhere in the universe) to suggest a cometary impact may have been responsible.

An impact in one of the oceans seems likely, but evidence for a crater on the sea floor could well have disappeared by now because the sea floor is completely replaced every 200 million years by subduction (the way the earth's tectonic plates slide under each other in the ocean beds). One possible site is off the northeast corner of Australia, and another is under the Antarctic ice cap. There are very large craters (250 km and 480 km [150 and 300 miles] in diameter, respectively) in both places that are formed by a ring of very high undersea mountains. Like the cometary impact that resulted in the extinction of the dinosaurs (see Question 47), this period is also associated with extensive vulcanism on the opposite side of the planet (in this case, the Siberian traps). Unusually, around 20% of the Siberian lava outpouring appears to have been pyroclastic (ash and acid aerosols thrown high into the atmosphere), and this would have added significantly to the nuclear winter effect created by the comet's impact.

The Triassic-Jurassic event (201 million years ago) resulted in the extinction of an entire class of sea creatures (the conodonts), 25–30% of all marine genera, and 42% of all terrestrial tetrapod genera (species that walk on four legs), all in a period of less than 10,000 years. Unusually, the fossil pollen record indicates that there was also a dramatic loss of plants, with around 60% of plant diversity being lost. The most likely explanation in this case is a period of intense volcanic activity: the Central Atlantic magmatic province (the huge area of volcanic outpourings down the middle of the Atlantic) occurred at exactly this time. It is the planet's largest extrusion of basalt lava flow, and may have been responsible for the breakup of the supercontinent Pangaea. It would have released massive quantities of carbon dioxide and sulfur dioxide into the atmosphere, causing climate warming.

That leaves us with the fifth mass extinction, the one associated with the disappearance of the dinosaurs.

47. Did the dinosaurs really go extinct?

The dinosaurs were the dominant life form on the planet for the better part of 300 million years (the Age of Reptiles). Then, quite suddenly around 66 million years ago, they all disappeared and were replaced by mammals. Quite suddenly is, of course, a relative term. It certainly didn't happen overnight. Rather, they gradually disappeared over a period of a few tens of thousands of years. The reason for their extinction was one of the iconic puzzles of geological science for the better part of a century.

Then around 1980, the father-and-son geological team of Luis and Walter Alvarez noticed that all over the world there was a thin dark geological stratum at exactly this date. This thin layer (barely a few centimeters [one inch] thick in most places) contained unusually high levels of iridium, a chemical element that is very rare on earth but abundant in asteroids. It was also full of shocked quartz and minute spherules of crystallized molten rock of the kind caused by cometary impacts. They suggested that the extinction of the dinosaurs had been caused by a massive comet or asteroid strike. This period is also associated with extensive volcanic activity, notably the Deccan traps (an area of lava flood 2,000 meters [6500 feet] thick with a total volume of a million cubic kilometers [c.233,000 cubic miles]) in India—almost as though the shock waves of the impact reverberated through the planet and emerged on the other side as a volcanic outpouring. The poisonous gases (in particular, sulfur dioxide) and ash released by this would have added significantly to the greenhouse effect produced by the gases and incandescent debris emitted by the impact of the comet on the other side of the world.

The clinching evidence came from the discovery of the 150-kilometer (94-mile) diameter, 20-kilometer (13-mile) deep Chicxulub crater under the southwestern Caribbean with Mexico's Yucatán peninsula forming its southern rim. It has been estimated that the asteroid must have been at least 10–15 kilometers (6–9 miles) in diameter. Evidence for tsunami beds along the Mexican coast and right up into Texas, as well as the much greater depth of the ash layer (up to a meter [39 in] thick, suggesting close proximity to the impact site), clinched the argument. The friction created by such a large object as it fell through the atmosphere would have set the atmosphere alight, and the impact itself would have thrown huge quantities of debris up into the high atmosphere. There is fossil evidence of instantaneous die-off and burial of animals as much as 5,000 kilometers (3000 miles) away. The resulting dust cloud would have obscured the sun over the whole of the planet, plunging the earth into darkness, and covered much of the ground with ash for at least a decade. Photosynthesis would have been impossible, and there is evidence for the extinction of 57% of all North American plant species. Incidentally, there is evidence of a line of smaller possible impact craters across the western hemisphere at the same latitude, suggesting that a much larger body may have broken up as it entered the earth's atmosphere. If so, this would simply have exacerbated the effects.

Everything that lived on vegetation would have very quickly starved, followed by the carnivorous species that ate the herbivores. It has been estimated that 75% or more of all species went extinct. Entire groups of previously successful species disappeared, including the dinosaurs, the plesiosaurs (marine dinosaurs), giant marine lizards, and ammonites (a family of giant mollusks related to the octopus whose curled-up shells could be as large as 2 meters [6 feet] in diameter—which, sliced and polished, are much prized today as decorative artwork). Seven out of 41 families of sharks, rays, and skates also disappeared at this point. The species that survived were mainly omnivores, insectivores, and carrion eaters, and small in size. The only reptile species that survived were ones weighing less than 25 kilograms (55pounds) (e.g., turtles, small crocodiles, and snakes).

One group of small dinosaurs did survive, however: they were the ancestors of our modern birds. So, in one sense, the dinosaurs never went extinct. Some of them are still happily with us! Once the climate had settled down again, the birds, along with the small mammals that had survived, underwent dramatic radiations to produce most of the major groups that we are familiar with today. Even so, mammals do not start to show any significance presence in the fossil record until around 185,000 years after the impact.

48. Have humans ever been responsible for extinctions?

Until a few thousand years ago, the human population was too small and scattered to have done more than local damage to habitats and their wildlife. Concentration of large numbers of people in urban centers beginning in the Neolithic certainly put pressure on the local environment as trees were cut for building materials and fuel, and fields began to be intensively cultivated to provide food to sustain a highly localized population, all of which inevitably resulted in erosion and desertification. The city of Akko on the Mediterranean coast of modern Israel (the Acre of the medieval Crusaders) has been continuously occupied for as long as 6000 years. Its archeological record details the dramatic impact that urbanization had on the local environment, changing it forever from a densely forested landscape to much less productive dry scrubland. Similarly, Easter Island in the southeastern Pacific was densely forested when it was first colonized by Polynesians around 1200 AD, or possibly a little earlier; but by 1722 when the first Europeans arrived, the island had been completely deforested and what had once been an economically and socially prosperous society had collapsed into famine and civil war.

It seems to be a commonplace that the arrival of modern humans has resulted in the collapse of the local animal communities, even in the absence of urbanization. This has often hit the giant species (the megafauna) hardest (for the reasons discussed in Question 46). The two best-known examples are the extinction of the Australian marsupial megafauna soon after the Australian Aboriginals arrived on the continent (around 40,000 years ago) and the disappearance of the North American megafauna soon after the arrival of the first wave of Native Americans around 12,700 years ago (when 90 genera of species that weighed more than 45 kilograms [100 pounds] disappeared abruptly, including giant sloths, short-faced bears, tapirs, peccaries, the American lion, saber-toothed cats, a llama and a camel, various large antelope and deer, horses, mammoths, mastodons, the giant armadillo, and giant beavers). Both seem to be cases of humans overhunting large, slow species that had not evolved any defenses against a mobile hunter with weapons, and which lacked the resilience to breed fast enough to offset the rate at which animals were being killed.

Other cases have involved individual species, in many cases on isolated islands, that have been driven to extinction by humans either through overhunting or through the deliberate or accidental introduction of pest species. Among the best-known examples are the dodo (a large flightless bird that was found only on the Indian Ocean island of Mauritius: it was hunted to extinction by the 1660s) and the passenger pigeon (once the most abundant bird in North America, but driven to extinction in 1901 by hunting). Until the 1850s, tens of millions of North American bison roamed the plains of the Midwest in huge herds, but settlers slaughtered an estimated 50 million over the course of the nineteenth century until only a handful were left and the species was on the brink of extinction.

Rats escaping from ships onto remote islands have been a particular problem, especially for ground-nesting birds that have no natural predators. Almost every oceanic island has suffered from this since Europeans began serious sea voyaging in the sixteenth century.

Tragic though these individual cases are, they pale into insignificance compared to the habitat destruction that has been inflicted on the planet in the past century or so. Vast tracts of virgin tropical forest have been cleared either for timber or to make way for plantations of commercial crops, ranging from rubber trees in the 1880s to oil palm in the 2000s. Orangutans, the last remaining Asian ape, have been reduced to a few thousand individuals, increasingly squeezed into a handful of (for the moment) protected areas.

Industrial activities have also had unintended consequences. The widespread use of DDT to control agricultural pest species after the Second World War started to have devastating consequences for local bird populations in North America because it caused eggshell thinning, and a resulting failure to reproduce. It was only when the biologist Rachel Carson alerted the world to this in her book *Silent Spring* that action was taken and the use of DDT banned. More recently, Indian white-rumped vulture populations experienced a similar problem, declining from 80 million in the 1980s to a few thousand by the turn of the millennium. It transpired that the cause was diclofenac, an anti-inflammatory drug administered to livestock that is fatal to vultures. The vultures had ingested it when feeding on the carcasses of animals that had been dosed with it.

The loss of habitats, wildlife, and plant species that has occurred within the last century has been 100–1,000 times greater than the background rate of extinction (see Question 45). One estimate suggests that as much as 7% of all species on the planet have been lost in the last two centuries alone. For this reason, it is being widely seen as the sixth mass extinction event, and the first to be caused by another species rather than by natural causes.

49. Can species rise from the dead?

The film *Jurassic Park* is based around the idea that it might be possible to recreate extinct species, in this case dinosaurs, by breeding them from their DNA. This wouldn't actually be possible for dinosaurs, because they died out too long ago for the DNA to have survived. Although DNA is a very tough molecule and quite resilient, it does eventually decay. The oldest DNA we have is only fragmentary (i.e., not a whole genome) from a horse that died 700,000 years ago. DNA has been successfully extracted from mastodons and Neanderthals that lived 50,000 years ago and a polar bear from 110,000 years ago. Even so, these DNA sequences are incomplete. Once the animal's skeleton has turned into completely solid rock, as is the case with older fossils, there isn't much hope of extracting anything.

A more promising possibility is reconstituting species that went extinct only relatively recently. The quagga was a member of the zebra family that lived in southern Africa but was hunted to extinction in the wild by 1878, with the last captive animal dying in 1883—recent enough that we have one photograph of what it looked like and about a dozen complete skins. It might be just possible to use cloning techniques to reconstitute a quagga by inserting DNA extracted from a skin into the egg of the Burchell's zebra (of which it may in fact have been a subspecies) and using a zebra as a mother. The same has been suggested for woolly mammoths using elephant mothers. A similar project has been considered for the passenger pigeon (see Question 48), using the closely related band-tailed pigeon.

There are a few cases of species apparently coming back from extinction. These are often called Lazarus species (after the person of that name in the New Testament whom Jesus raised from the dead). However, all of them are in fact instances in which we *thought* the species had gone extinct but then a few individuals turned up alive. The most famous case is undoubtedly the coelacanth, the large (two-meter [6-foot] long) lobe-finned deepwater fish thought to have gone extinct 68 million years ago. They are relatives of the lungfish rather than the family of ray-finned fishes to which most modern fish species belong.[1] In 1938, one that had been captured by a native fisherman off the coast of South Africa was identified by a marine biologist. In fact, these fish may not have been that unfamiliar to the fisherman; however, because coelacanths are adapted to life at depths of around 500 meters (1600 feet), they decay rather fast when brought to the surface and so were usually thrown back as inedible. Quite a number have since been caught and examined by biologists. They are all but identical to their ancestors from 400 million years ago, another example of how life in the ocean depths is buffered against the selection pressures of climate change (see Question 16).

Another possible case is the marsupial Tasmanian tiger, or thylacine. Once widespread as the top predator throughout Australasia as far as New Guinea (it appears in Australian rock art from 1,000 years ago), it had gone extinct on the Australian mainland by the time Europeans arrived in the 1700s. It had survived only on the island of Tasmania but was thought to have gone extinct even there by the 1930s. During the 1990s and 2000s, however, a number of observations and photographs purporting to be thylacines were recorded from remote areas in Tasmania, although so far none have been confirmed and no animals have actually been caught.

50. Does a species' genetic diversity matter?

We often make desperate attempts to save the last few remaining populations, or even individuals, of species that are on the verge of extinction. These are important last-ditch conservation efforts that we often support enthusiastically with our donations. However, simply because we have saved a few hundred individuals does not mean that the species is saved and will thereby survive forever.

Small populations inevitably suffer from founder effects (see Question 43) because the individuals concerned represent only a small fraction of the species' original genetic diversity. Lack of genetic diversity means that a population does not have the inherent flexibility to respond to new conditions (see Question 45). It also makes it more likely that the surviving animals will suffer from genetic or physical disorders that impair their survival—something we are familiar with from the fact that many inbred dog breeds suffer from a variety of problems. Loss of genetic diversity is likely to make a species vulnerable to diseases because it doesn't have the scope for generating immunities. It may also limit a species' ability to evolve the capacity to cope with environmental stresses.

We see this demonstrated especially clearly in domestic crops, which have often been so selectively bred over thousands of generations that they have very little genetic diversity left. As a result, commercial tobacco and coffee crops were devastated by fungal diseases at various times during the nineteenth and twentieth centuries. During the 1870s, the French wine industry was virtually wiped out by grape phylloxera (a microscopic sap-sucking insect related to aphids), for which there was no chemical defense; the French vineyards were saved only by the ignominy of having to import phylloxera-resistant American vines.

The effect of population size on diversity has been confirmed in a recent experiment using a virus as a parasite and a bacterium as the host, with host populations being manipulated to have more or fewer clones (and hence more or less genetic diversity). Viruses can evolve rapidly in a very short time to overcome the defense mechanisms of their hosts, so the experiment asked whether populations composed of more clones were less likely to be taken over by the parasite because the parasite was unable to adapt to many different defenses simultaneously. The experiment confirmed that this was indeed so.

A particularly instructive example of some of the problems that emerge from reduced biological diversity (as well as some of the consequences of humans interfering in habitats) (see Question 48) is provided by the Green Revolution of the 1960s. New breeding techniques were used to develop novel, spectacularly high-yield varieties of cereals like wheat and rice. This was credited with saving a billion people from starvation, and earned its initiator, Norman Borlaug, a Nobel Prize. But the intensive breeding programs involved inevitably resulted in loss of the genetic diversity that had been bred into traditional varieties of wheat and rice over many thousands of years.

As a result, the new commercial varieties were more susceptible to local parasites and plant diseases, and thus they invariably required the use of large quantities of insecticide, as well as nitrate-based fertilizers because the new varieties were very demanding of the habitats where they were used. In the Punjab, which pioneered the implementation of the Green Revolution, the consequence was that, by 2009, wells had nitrate levels that were 20% above WHO-recommended levels for safe drinking water. The region also had cancer rates (attributed to the overuse of pesticides) that were significantly above previous levels. The much-vaunted Green Revolution turned out not to be so cost-free.

6

EVOLUTION OF COMPLEXITY

51. Why do some species eat others?

Life began as microscopic single-celled organisms making
their own energy from sunlight and other chemical resources.
At some point, some of these creatures found it less effort just
to eat their neighbors and acquire the energy they had gone to
such great efforts to create. With that, a complex hierarchy of
predator–prey relationships was created. One of the defining
features of such hierarchies, which still holds good today, is
that the organisms that fill each successive layer will generally
be larger, but fewer in number, than those in the layer below
them. Larger animals also have lower birth rates because they
require more investment to produce—a phenomenon recognized as long ago as 350 BC by the great Greek philosopher-
scientist Aristotle.

An important phase transition occurred with the evolution
of plants around 800 million years ago (see Question 32) because it created a whole new unexploited resource for any
species that was capable of exploiting them. However, early
plants (mainly ferns and conifers) were not especially succu-
lent or nutritious, and it was not until 160 million years ago
the really big development occurred following the evolution
of flowering plants (the angiosperms) and in due course grasslands (see Question 34). This led to two major developments.

First, it spawned an adaptive radiation (or explosion of speciation events) of grazers and browsers adapted to exploiting this new foraging niche. Second, it gave rise to a new class of predators adapted to feeding on these new herbivores. I suppose we might actually add a third wave in which the plants turned on some of their predators and ate them—though this is mainly confined to insectivory (the pitcher plants, sundews and butterworts, and Venus flytraps, among many others that feed on insects).

Thus, two major types of ecological relationships—plant–herbivore interactions and predator–prey interactions—emerged in very quick succession. These have shaped the evolution of ecosystems ever since. In fact, both kinds of interaction are functionally very similar. They involve a predator that exploits a prey. The only real difference is that in one case the prey are plants and static, while in the other they are animals and mobile. But aside from these minor details, much the same principles apply to their interactions. In both cases, prey species will evolve strategies designed to outwit predators in order to avoid being eaten, and predators in turn respond by evolving strategies to circumvent these.

A very nice example of this is provided by the way brain size increases through evolutionary time in the carnivorous mammals and their ungulate prey. Carnivore brain size increases first in order to hunt mobile prey more effectively; ungulate brain size then increases so they can outwit the predators; then carnivore brain size increases again to outwit the outwitters, after which ungulate brain size increases a bit more. It is a classic example of an *evolutionary arms race*, where the stakes are progressively ramped up over time as species (or, rather, individuals) try to outcompete each other.

52. In what ways do plants and animals exploit each other?

Plants and the herbivores that eat them have engaged in a particularly remarkable evolutionary dance. Plants have evolved a number of strategies to avoid being eaten, or at least to be eaten only when it suits them to be eaten. Some of these are mechanical (the development of thorns to deter herbivores, hard shells to protect their seeds) and some are chemical (poisons). Many plants, for example, protect their fruits inside hard cases, which split open naturally to release the seed when it is ready to germinate.

Some plant species want their seeds to be taken as far away as possible so that the seedlings do not compete with each other as they grow. When they are ready to germinate, these species make their seeds as enticing as possible by covering them in a nice, sweet, fleshy coating that will encourage herbivores to eat them. The sugary flesh provides the herbivore with energy in exchange for carrying the seeds away in their gut; a day or so later, the seeds will be excreted some distance away where they can now germinate. In fact, the seeds of some species will not germinate at all if they have not first passed through a herbivore's gut: the acids in the stomach help prepare the seed for germination (for example, by weakening the casing so that it will break open and allow the seedling to emerge).

If herbivores eat the fruits before the seeds have had time to develop properly, the parent plant will have wasted an enormous amount of effort. To discourage herbivores from eating their fruits until they are ready to germinate, plants infuse the flesh with secondary compounds that herbivores find distasteful. These can include alkaloids (bitter compounds that actively block the digestive enzymes and make you sick: these include nicotine, cocaine, strychnine, and cyanide), terpenoids (like citronella, menthol, and pinene, some of which give rise to dangerous glysosides like digitalis), or phenolics (like cannabinoids that disrupt endocrine function or tannins that interfere with the digestion of protein). As the fruits ripen and the seeds become ready to germinate, these poisons naturally decay and are replaced by sugars to attract the herbivores.

Inevitably, some herbivores have developed tolerances to some (but never all) of these so that they can eat unripe fruit.

Apes cannot detoxify the tannins—which is why we get that sudden drying in the mouth (and, as our mothers always warned us, a stomach ache) when we eat unripe fruit. But Old World monkeys evolved the capacity to detoxify the tannins in unripe fruits, which gave them an edge over the apes (who were at the time the most abundant primates).

Fruit-eating birds and mammals play an especially important—if unintended—role by acting as seed dispersers. But we shouldn't forget all the insects that act as inseminators for plants by conveying pollen from one flower to another as they go about their business looking for the nectar that the plants kindly provide as inducements. Many plants are, of course, pollinated by the air, but that is a bit hit-and-miss. So enticing a helpful insect like a bee to carry your pollen directly to your reproductive partner is much more efficient.

These kinds of interrelationships between animals and plants can become immensely complex, creating the "entangled bank" that Darwin referred to (see *Preface*). They also give rise to a concept of central importance in ecology, the *food web*.

53. What is a food web?

One reason why species extinctions seem to go in clusters is that groups of species at a particular location are intimately linked in a set of complex relationships known as a *food web*. The food web is the set of who-eats-who relationships that exist between all the invertebrate, plant, and vertebrate species in a habitat. Like spider webs, food webs can be extremely complicated and intricate. Remove one species, and it immediately affects all the species that depend directly on it for their food, whether these are herbivores, carnivores, or detritus feeders. This in turn affects the species that depend on them, and the result can quickly turn into a domino sequence of collapses.

The standard structure of a food web involves three main layers or *trophic levels*: producers (plants and some microorganisms that create their own energy from sunlight), primary consumers that eat the producers (usually herbivores), and secondary consumers that eat the primary consumers (mainly carnivores). The last category can have several layers of its own if there are carnivores that eat other carnivores (polar bears that eat seals that eat fish), which is why they are usually referred to as *top predators*. These layers reflect the flow of energy up through the system, with a separate layer of decomposers taking the nutrients locked up in each level back to the base for use by autotrophs after organisms have died.

Autotrophs create energy out of sunlight and combine this with chemicals extracted out of the air (e.g., carbon from carbon dioxide) or the ground (e.g., the trace elements like iron, copper, magnesium, and zinc that are essential for life). Photosynthesis is possible because the pigment chlorophyll (which gives plants their green color) has the capacity to absorb energy from the blue and red portions of sunlight. Primary consumers eat the producers, and in turn provide a food source for the secondary consumers. The result is called a *food chain* because energy and other nutrients flow up it from producers to primary consumers and eventually top carnivores.

Because energy is wasted at each level (some energy is locked up in the cell walls of plants or the bones of herbivores and so doesn't get eaten, for example), each layer is always numerically smaller than the one it feeds on. Conventionally, the size of each layer is measured in terms of its *biomass*—literally, the weight (or, more technically, mass) of all the organisms in the layer added together. This is because body mass is a good proxy measure for the energy content. The result is an *ecological pyramid* that has a large bottom layer (the producers), a medium-sized middle layer (the various primary consumers), a small layer of secondary consumers, and a very small top carnivore layer.

The layers exist in a kind of dynamic equilibrium with each other. If there were too many primary consumers, they would wipe out the producers and the whole pyramid would collapse. Predators (say, lions) limit the number of herbivores by feeding on them, and this prevents overgrazing and helps to promote plant growth. In some cases, the feeding habits of a species may even facilitate the productivity of another species. Ranging goats with cattle on ranches, for example, allows cattle to be stocked at much higher densities because the goats eat out the rank vegetation that cattle cannot cope with, allowing grasses that would otherwise be suppressed by this vegetation to proliferate; this allows more cattle to be pastured.

These complex interactions lie at the heart of an ecosystem—the collection of plants, animals, decomposers, and other life forms that make up the set of species in a given location. These web structures can be very complicated, reflecting a complex optimization between the interests of the different species. Removing one species can have dramatic and unpredictable effects. Krill are six-centimeter (3-inch)-long shrimp-like crustaceans that abound in the southern oceans around Antarctica (in particular). They feed on microscopic phytoplankton, and in turn provide food for fish, penguins, and baleen whales (which use the whalebone plates in their mouths to filter out huge quantities of krill); they, in their turn, variously support seabird, seal, and killer whale populations.

Before humans started hunting them in industrial quantities in the nineteenth century, baleen whales consumed huge quantities of krill—probably as much as half the 400 million tonnes/tons of krill present in the Southern Ocean—and were responsible for recycling much of the iron locked up in the krill and making it available for phytoplankton in an important feedback process the underpinned the Southern Ocean food chain. Overhunting of the whale populations in the twentieth century initially resulted in rapid increases in the penguin and seal populations because there was so much extra krill around. But later on, the breaking of the whale feedback loop resulted in a decline in krill numbers when they became starved of the nutrients previously recycled by the whales, with consequent declines in the penguin and seal populations (although climate change may also have played some role in this).

54. Do ecosystems evolve?

Given that species evolve, and even that groups of species can coevolve (for example as predators and prey) (see Questions 52 and 53), it is perhaps natural to ask whether ecosystems that consist of many species bound together in a complex web of interrelationships also evolve. If evolution just means change over time, then of course they do, and that probably isn't terribly interesting. What is more interesting is whether there are regularities in their evolutionary trajectories. After all, the universe itself has evolved, and in a very prescriptive way driven by the fundamental laws of physics. But does this also apply to environments and their constituent animal and plant communities?

At one level, the answer is definitely yes (see Question 53). But there are other regularities in the way ecosystems as a whole evolve. If forests are destroyed by volcanic eruptions, hurricanes, wildfires, or even glaciations, recolonization follows a very regular pattern that may in fact reflect the evolutionary history of plant life and terrestrial habitats. Initially, in a phase known as primary succession, algae, lichens, and fungi colonize the area and help to build up the soil (they are particularly important if the destruction has removed most of the soil, as glaciations often do). These are referred to as pioneer species because they appear first. As soil quality improves, species like grasses and ragweeds that can cope with poor soils appear. These help to stabilize the soil by preventing wind erosion; they also help to hold water in the soil, a crucial step for the next phase. This phase is dominated by species with small seeds or spores that can remain dormant for long periods if the habitat dries out but which can also easily disperse, especially on the wind. These species are often ones that can fix nitrogen in the soil by extracting it from the air. This is crucial for the next phase as nitrogen is often a limiting resource for plants, most of which acquire their nitrogen from the ground. As they die and decay, they build up soil depth and richness.

The second phase (secondary succession) involves the arrival of more conventional shrubs and then trees, often in a well-defined sequence. Annual plants are often the first to appear, as they can cope better with poor, uncertain conditions. After these have established themselves, shrubs and small trees like pines, oaks, and hickory (known as "colonizer species") become established. Once these get established, they are gradually replaced by more substantial forest trees that introduce more canopy layers. These tree species often fight for sunlight rather than soil, which is why they typically grow very tall and why there is often very little ground-level vegetation beneath them. After about 150 years, the forest will have a species composition that will remain stable, often for many thousands of years. These are known as "climax habitats." Of course, the composition of species may not be exactly the same as the original forest, since it will depend on accidents of history as to exactly which species are present in potential source habitats nearby and which of these are then actually brought in by seed dispersers like birds and herbivores.

The exact composition of the climax habitat in a particular location ultimately depends on the local soils and climate. A dry habitat may never be more than thornscrub-dominated woodland or, if drier still, just grassland. And even a patch of dense forest may never return to that state if the habitat has become so degraded that its geological structure has changed forever (as seems to have happened in many Mediterranean habitats after the Neolithic resulted in dense human settlement) (see Question 48). The broad sequence, however, is relatively stable. The stability of the sequence also reminds us that it is the competitive advantages of different species (i.e., their adaptations) that determine who arrives when and how successful they are at the different stages.

Animals may also colonize such habitats in a particular sequence. Early colonizers will usually be insects, and then perhaps birds—species that can come and go over larger areas if things get tough for them locally. Later, burrowing species such as insectivores (e.g., voles) or rodents (e.g., mice), amphibians, and reptiles (e.g., snakes) will colonize. Only once the habitat has developed sufficiently will large herbivores be able to survive there, and only once they have become established will the larger predators that depend on them return.

55. Does it matter if we upset the "balance of nature"?

Ecosystems evolve over many millennia and, providing circumstances do not change dramatically, will eventually arrive at a reasonably stable state—the "balance of nature." However, because they often involve a great many different components (i.e., species), ecosystems are often very finely balanced. Disrupt one part of the chain, and it may ricochet down through the system in what is known as a *trophic cascade*. We humans often intervene for reasons that seem quite sensible at the time, but because we do not understand the full complexity of the local ecosystem, our interventions can have unintended consequences that may be hard to undo.

The most famous case of this is probably the introduction of rabbits to Australia, seemingly just to have our European furry friends in their gardens. At the time, this no doubt seemed an innocent enough thing to do. In fact it proved to be more difficult than people had imagined. It took as many as a dozen attempts for them to finally get established. But once they did become established, their populations exploded and, as grazers, they had a terrible effect on the fragile grasslands of the hinterland. They have proved impossible to remove. Another infamous case was the use of DDT as an agricultural pesticide (see Question 40).

A more recent example was the introduction of two species of planarians (flatworms) into the Philippines, Indonesia, New Guinea, Hawaii, and Guam to control populations of giant African snails that had previously been released in these islands and were displacing native snails. It worked very well initially, and seemed to provide an exemplary case of biological pest control. However, it seems that having exhausted the supply of giant snails, these parasites are now attacking the native snails. So the solution has merely reinstated the original problem in a different form that is much less easy to deal with.

The removal of top predators from habitats often creates unintended problems. One historical example involved the sea otters of the Aleutian Islands off Alaska. When the nineteenth-century fur trade wiped out the otter populations, the sea urchins on which they were the main predators exploded and started to eat their way through the kelp forests in the waters around the islands. As a result, the fish populations that had used the kelp both as refuges from fish predators and as rich food sources also went into decline. The bald eagles that had lived on the fish were in turn forced to change their feeding habits, impacting other prey species. Similarly, when wolves were eradicated from Yellowstone National Park in the United States, it resulted in an increase in the elk populations, who then grazed and trampled increasing numbers of aspen seedlings, so aspens started to become less common.

The number of examples multiply with every decade. In large part, it is simply a consequence of the fact that our understanding of the relationships between the various species in a habitat is lamentably poor. It is also partly because our own cognitive limitations make it difficult for us to think in more than very simple cause-and-effect terms. Moreover, it can sometimes take a long time for these effects to work themselves through and become apparent so we don't spot the problem until it is far too late. One example of that is provided by the overhunting of baleen whales in the Southern Ocean during the late nineteenth and early twentieth centuries (see Question 53). It took half a century for the full impact to become apparent. Just because everything seems fine when we first intervene does not mean that there won't be any consequences later on. The lesson is that we interfere with these fine-tuned evolutionary processes at our peril.

56. Why did sexual reproduction evolve?

Sexual reproduction is, in many ways, the archetypal example of evolutionary complexity. But it remains one of the least well-understood aspects of organismic biology. The earliest forms of life were prokaryotes—single-celled organisms like bacteria that reproduce by simple division (see Question 32). Some of these prokaryotes exchange genetic material by pairing up (a process known as *conjugation*), which differs from sexual reproduction in that the gametes exchanged do not differ and gamete exchange may in any case be incomplete. In contrast, almost all eukaryotes reproduce sexually, although some, like the humble garden aphid, can alternate between sexual and asexual reproduction. Sexual reproduction probably evolved only about a billion years ago.

One problem with sexual reproduction is that we gratuitously throw away half our genes during the process leading to fertilization. In contrast, asexual reproducers get a full complement of their genes into each "offspring" every time. In addition, multicellular organisms have the added disadvantage that their greater complexity means that individuals develop more slowly. This longer generation time means that they cannot respond so quickly to new parasites and diseases. This constitutes the *twofold cost of sex*, and it has puzzled evolutionary biologists for the better part of a century. Why would it ever be worth evolving sexual reproduction?

One suggestion is that it allows the organism to cope with a wider range of parasites. The relationship between hosts and parasites is a bit of a *Red Queen process*—named after the Red Queen in the *Through the Looking-Glass* sequel to Lewis Carroll's *Alice in Wonderland.* Just as the Red Queen had to keep running in order to stand still, so organisms have to keep adapting in order to keep up with the fast mutation rates of parasites. Each individual's immune system will have been exposed only to a limited range of parasites, so by mixing genes with another individual who has experienced a different array of parasites the organism will increase the range of parasites it can cope with. This was demonstrated in one recent experiment in which the roundworm, C. *elegans* (whose mode of reproduction can be switched genetically between sexual reproduction and self-fertilization), was exposed to a fast-evolving pathogenic bacterium parasite. The results confirmed that self-fertilizing populations were indeed more likely to go extinct as the parasite evolved than populations that reproduced sexually.

Sexual reproduction might fortuitously allow new advantageous genotypes to appear with entirely novel adaptations. Another possibility is that it allows deleterious genes to be hidden (through Mendel's dominant/recessive mechanism) (see Question 22) or even eliminated completely. By separating the two halves of the chromosome during meiosis, the deleterious mutant allele is separated from the normal allele, allowing the normal allele to reproduce. That way, you lose only half of your offspring instead of all of them. In effect, sexual reproduction acts as a natural filter to weed out deleterious mutations.

One intriguing question is why there are only ever two sexes. Although in principle it would be perfectly possible to have many different types of gametes (and hence many different sexes), in fact no plant, animal, fungus, or even those protists that have sexual reproduction (e.g., the malaria plasmodium) ever have more than two sexes. There are occasional hermaphrodites (see Question 57), which appear to be mainly due to developmental errors, but there are no natural cases of three or more sexes. To investigate why this might be so, evolutionary biologists have had to resort to mathematical modeling to explore the question; the answer seems to be that populations with more than two sexes are unstable.

57. How are the two sexes determined?

One of the universal features of sexual reproduction is that one sex produces a much larger gamete (usually called an egg) than the other (which produces sperm), and hence pays a much higher price for reproduction. Usually, this is because the egg incorporates a large amount of cytoplasmic material in addition to the nucleus and its chromosomes, whereas the sperm is just a nucleus shorn of any extras. There is no entirely satisfactory explanation for why this should have evolved, other than the fact that it may well be the result of an evolutionary arms race in which one sex cheated on the other by progressively reducing its investment in order to force the other to carry the greater cost. The other sex is then caught in a bind in which it is offered the choice between investing more heavily so as to reproduce or not reproduce at all—sometimes known as a *Hobson's choice* strategy.[1]

We tend to call the sex that produces the egg a female and the one that produces the sperm a male, but that is mainly because this is what happens in mammals. In mammals, the two sexes are differentiated by their sex chromosomes: females have two normal-sized X chromosomes, whereas males have one X chromosome and a tiny Y chromosome to pair up with it. The human Y chromosome has only 59 million base pairs with just 70 protein-coding genes, compared to about 155 million base pairs on the X chromosome with some 800 genes. The Y chromosome appears to have once been a perfectly normal-length chromosome similar to the X but to have lost most of its genetic material during the course of its evolution.

The Y chromosome seems to do very little aside from switching the fetal brain over to a male form. All mammal fetuses initially have a female form, but those with a Y chromosome switch the fetal brain into male format, and the resulting testosterone surge at puberty will then switch the body into male body form. This is not entirely a matter of genes. The trigger for this in the embryo is actually the speed at which the embryo grows (male embryos grow faster, a phenomenon known as the "race to be male"). Because there is a developmental trigger and this can depend on circumstances (for example, how well fed the mother is), some male embryos fail to make the switch and are born as females. This occurs with a frequency of about 1 in 15,000 female births. These individuals often have a more androgynous (male-like) body form, are often good athletes, and usually have an advantage when competing against chromosomally normal women. At the 1996 Summer Olympics in Atlanta, eight women athletes were found to be chromosomally male (i.e., XY). In 1950 the Dutch 200-meter world record holder Foekje Dillema was banned for life from competing in international athletics because she refused to take a genetic sex test when these were introduced. Subsequent tests indicated that although she was phenotypically female, genetically she was an XX/XY mosaic (some of her cells were XX and some XY). It seems that this probably developed from an XXY genotype.

In some cases, accidents of inheritance can give rise to XO individuals in humans (usually because one of the X chromosomes is missing or fragmented). Known as Turner syndrome, it occurs in around 1 in 5,000 women. It is characterized by short stature, a webbed neck, and low-set ears; these individuals are usually sterile, and prone to obesity, as well as heart and thyroid defects. Some males can end up with extra X chromosomes (XXY, XXXY, or very occasionally even XXXXY), a condition known as Klinefelter syndrome; the symptoms are often unspecific but can include unusual stature, poor coordination, small testes, and disinterest in sex, with these being worse the more X chromosomes the individual has. XXY occurs with a frequency of 1 in 1,000 male births, XXXY in 1 in 50,000 births, and XXXXY in 1 in 100,000 births. The cause is not genetic, but beyond that it is poorly understood. XYY males also have a frequency of 1 in 1,000 births, but aside from being taller than average and having an elevated risk of learning difficulties, most individuals are normal. There were early claims that XYY individuals were more prone to aggression, but this seems not to have been borne out when larger samples were examined. Again, the cause seems to be an accidental duplication of the Y chromosome during meiosis.

During the 1930s, British geneticist Ronald Fisher showed that frequency-dependent selection should result in equal numbers of the two sexes in the reproductive age cohort. This is because if by chance the frequency of one sex increases, it will automatically make the rarer sex more valuable and natural selection will quickly favor mechanisms that produce more of that sex. However, he also pointed out that this argument really applies to the *investment* in the two sexes, not to the simple number of the two sexes. The cheaper sex is usually the males, and in humans around 140 males are conceived for every 100 females. Partly because they have only one X chromosome, males experience more developmental disorders, such that by the time they are born, this number has been whittled down to about 108 males for every 100 females. Males continue to have higher mortality rates, both during infancy (mainly due to greater susceptibility to respiratory and other childhood diseases) and during the teenage years (due to higher risk taking); by the time they are ready to reproduce, the ratio finally approximates equality.

58. Do all organisms determine their sexes in the same way?

It seems that in concocting two sexes, evolution has had some fun at our expense. There are more ways that two sexes are produced than there are human marriage systems. While mammals all have an XX/XY chromosome system, birds, some fish and crustaceans, as well as butterflies and moths, arranges things the other way around: the XX sex produces the sperm and the XY sex the eggs. Because different chromosomes serve as the sex chromosomes, the bird version is usually referred to as the ZW system. As in the mammalian system, the Z chromosome is larger and has more genes. There are no genes in common between the XY chromosomes of mammals and the ZW chromosomes of birds, so it seems that the sex chromosomes in each case evolved independently from different autosomal chromosomes. It can, however, get worse: the platy fish combines both systems, having W, X, and Y chromosomes; nonetheless, it still has only two phenotypic sexes, with females being WX, WY, or XX while males are XY or YY. They seem not be too bothered by this.

Some groups of animals have only X chromosomes, with males having only one chromosome (hence XO) while females have the usual two. This system has been adopted by spiders and scorpions, dragonflies and silverfish, grasshoppers, crickets and cockroaches, and some nematode worms, crustaceans, and mollusks, as well as a few families of bony fish. In other groups (notably bees, wasps, ants, and thrips), sex is determined by whether or not an egg is fertilized by a male. In these species, males come from unfertilized eggs and have only one set of chromosomes, whereas females come from fertilized eggs and have the usual two (a genetic system known as *haplodiploidy*). That has the odd effect of meaning that a male has no father and can have no sons, but can have both a grandfather and grandsons. It also means that females are more closely related to their sisters than to their own offspring.

In some groups of animals, sex is determined by circumstances. In tropical clownfish and the wrasse families, everyone is born and lives as a female; whenever the single male in the group dies, the dominant female transmutes into a fully functional male and takes over as the local male. In reptiles and teleost (or ray-finned fish), the sex of offspring is determined by the temperature of the nest where the eggs develop. In turtles, eggs are more likely to hatch into males if the temperature is low and females if it is high, but the reverse is the case in crocodiles and alligators.

My personal favorite, however, is the Mediterranean Bonellia worm. This humble walnut-sized creature begins life as a tiny leaflike larva released by its mother. It floats about in the sea until one of two things happens. If it finds a suitable substrate, it attaches itself and develops into a walnut-sized female; if it happens to get swallowed by a female as it floats by, it becomes a male and, along with varying numbers of other males, lives inside the female, fertilizing her eggs from within.

Another strange example, and one that perhaps tells us something about how two sexes evolved (see Question 51), is the flatworm (one of the platyhelminth family that includes tapeworms, lung and liver flukes, and the schistosomes that cause bilharzia). These are hermaphrodites and during mating they fence with a pair of dagger-like "penises" for up to an hour. The winner is the one that manages to cut through the partner's epidermis and inject sperm into its cardiovascular system. Since this fertilizes the loser's eggs, it forces the loser to become the "mother." How weird is that?

59. Why is there a trade-off between reproduction and parental solicitude?

Fitness (contributing genes to the species' future gene pool) (see Question 25) is not just a matter of surviving longer or producing more babies. It is about surviving long enough to produce enough babies to rear some of them to adulthood so that they can reproduce in their turn. The trade-off between survival and reproduction is one decision a species has to make, but after that it has to consider a second trade-off—that between reproduction and parental investment (being sufficiently solicitous of your offspring to invest time and effort in them to enable them to develop into effective adults).

The trade-off between producing babies and rearing them to adulthood turns out to be a very general issue in biology. The overarching contrast is between species that are *semelparous* (and produce all their offspring at one go) and those that are *iteroparous* (and produce their offspring a few at a time over many successive birth events). The first, sometimes known as the "big bang" strategy, typically involves producing a vast number of eggs at one go and dumping them into the environment (usually a watery one), leaving the babies to fend for themselves. In most of these cases, the parents die quite soon afterward from the sheer effort involved. Salmon are the classic example. The idea is that you overwhelm the many predators that might eat your young by producing far more than they can possibly cope with. If only a tiny proportion make it to adulthood, it's fine: a tiny proportion of a million eggs is still a large number of surviving adults. Iteroparous species pursue the opposite strategy. They opt to produce just a few offspring at any one time so that they can invest in each one in order to maximize its chances of making it through to adulthood. Monkeys and apes (and, of course, humans) represent the limiting case of this: they typically produce just one baby at a time and invest very heavily in it (often over several years). Which strategy is best depends entirely on the circumstances you happen to face as a species. There is no best way of doing it!

The big issue that separates these two alternatives is brain size. Large brains allow behavioral flexibility, and that means they cannot be preprogrammed in what they do: how to adapt one's behavior to changing circumstances has to be learned and practiced, and that takes a lot of time and experience, and needs a big brain capable of learning. Because brain tissue can only be laid down at a constant rate, large brains require a long period for their development, and that means the parent(s) have to continue investing long after birth. So, in the limit, not only do the parents have to feed their growing young, as many birds and mammals do, but they may also need to help their offspring learn how to find food and cope with the complexities of social life.

In most species, investment in offspring ends with weaning. In primates, it continues until at least puberty because the processes and experiences of socialization play a crucial role in adult behavior (see Question 84). This period extends well beyond puberty in humans, mainly because we don't acquire full adult social competency until around our mid-20s when the brain finally stabilizes. Humans are even able to extend parental investment well beyond this into the next generation, in part through cultural inheritance of information (rules of behavior that give us shortcuts on how to cope with the complexities of the world we live in, especially the social world) (see Question 91) and in part through the inheritance of resources and wealth.

Decisions about parental investment (or parental solicitude) can become even more complex when parents actively bias their investment in favor of some offspring. This can happen if offspring of one sex are more valuable than offspring of another sex (female offspring may have more opportunities to reproduce than male offspring in some contexts, but the reverse may hold in other contexts); birth order may also be important, or it may even come down to the personal qualities of individual offspring. Decisions to invest differentially in offspring in this way will hold despite the fact that natural selection has instilled in all of us a very strong motivation to nurture each and every one of our offspring as they come along (see Question 7).

60. Why does the nature/nurture controversy continue to bedevil discussion?

The nature/nurture controversy was (and remains) a rather peculiar dispute, but it is instructive as an example of how easy it is to confuse biological issues. Ever since the mechanisms of genetic inheritance were discovered, there has been intense interest in how much of an organism's biology and psychology are due to its genetics and how much to the influence of the environment or learning. These were not entirely new questions. We have always known that children resemble their parents, both physically and behaviorally. Indeed, the careful breeding experiments of the Agricultural Revolution of the late eighteenth century had yielded dramatic improvements in the productivity of both crops and domestic stock precisely because this is the case.

The debate came to a head during the 1930s in the form of a clash between comparative psychologists (mostly in the United States) and the emerging discipline of ethology (mostly in Europe). The psychologists, notably the founding fathers of behaviorism such as John B. Watson and B. F. Skinner, as well as some social scientists interested in human culture like Franz Boas, believed that everything an animal (or human) did was learned; others, like the ethologists Konrad Lorenz and Niko Tinbergen, believed that at least some aspects of behavior were instinctive (and hence, by implication, genetically determined). This debate rumbled on for half a century, with no sign of any conclusion.

During the 1960s, however, biologists came to the conclusion that almost nothing is wholly genetically inherited or wholly the outcome of environmental experiences (or learning). Of course, some traits (such as eye color, color blindness, and certain developmental disorders like Down syndrome and Turner syndrome) have very simple genetic bases, while other traits (such as the religion one belongs to) seem to be wholly culturally inherited. However, most biological and psychological traits, like stature and intelligence, exhibit a mixture of genetic and environmental effects, with the rearing environment itself often influencing how the genes express themselves. You may have the genes to be tall, but growing up in a poor nutritional environment will stunt your growth. In part, this is because many hundreds of different genes are usually involved, each having a small additive impact (a topic known as *quantitative genetics*). Asking whether something is more or less genetically determined, or more or less learned, is a meaningless question, because it is impossible to separate out these two components in the developmental process (see Question 78).

Part of the problem may be a misunderstanding over the meaning of the technical genetics term *heritability*, which measures the proportion of variance in a trait that is due to genetics as opposed to local environmental conditions. This is commonly misinterpreted as meaning the proportion of a trait that is determined genetically. In fact, it refers only to the observed differences in a trait between individuals in a population, not to what determines the trait itself. Paradoxically, when all individuals have equal access to environmental resources and socialization experiences, the heritability is not 0% but 100% because the only variation left in the system will be that due to genes.

The origins of the dispute seem to lie in an implicit and well-meaning fear on the part of some psychologists and social scientists that if human intelligence or behavior has a genetic component, however small, then any prospect of improving the human condition is dead in the water. Alas, this is rather reminiscent of what became known as the Lysenko Affair. During the 1930s and 1940s, a little-known Russian biologist named Trofim Lysenko managed to gain the ear of Stalin and persuade him that modern Mendelian genetics (indeed, any kind of genetics) was anti-Marxist because it implied that it would be impossible to improve the human condition by implementing social reforms. Instead, he advocated something close to Lamarck's theory of the inheritance of acquired characteristics (see Question 3), claiming that poor-grade rye could be transmuted into high-grade wheat and barley by stressing the plants, and that agricultural productivity could be increased by planting the plants much closer together because, like good socialists, they liked to cooperate rather than compete with each other like wicked capitalists. The result was the death from starvation of more than a million peasants on the collective farms of Russia and China when these were forced to implement his theories during the 1950s and 1960s. This sad affair wasn't helped by the fact that most of the leading Russian geneticists of the prewar period had been executed or sent to the gulags. In the 1920s, Russian genetics had been decades ahead of the West, but that knowledge had been sidelined by ideology.

That the nature/nurture debate continues to hover omi-
nously over any discussion about human behavior probably
says more about our inability to grapple with complex biological processes and the fact that it is all too easy for politics to
seep into scientific debate and derail it.

7

EVOLUTION OF HUMANS

61. Who were our earliest ancestors?

Our lineage can be said to have started when the African great apes of the late Miocene gave rise to a new, more terrestrial group, the australopithecines, some six or seven million years ago. Two important things had happened at this point. One was that at least some of the Miocene ape species in Africa had decided to move down from the trees to pursue a more terrestrial life. Their descendants, the chimpanzees and gorillas, still share that trait with us. This is not to say that they cannot, or do not, climb about in trees—they simply became more terrestrial than their predecessors. This was probably a consequence of competition with monkeys for the dwindling food sources in tropical forests and the evolution of a dietary mutation that allows them to exploit the rotting fruits on the forest floor (see Question 20).

The second key event was that these more terrestrial apes gave rise to a lineage that began to travel bipedally so that they could access rich underground food sources on the floodplains of large rivers and lakes beyond the forest edge. Australopithecines were, nonetheless, apes in all but name, differing from the other African apes only in their bipedalism. Their brains were no bigger than the brains of contemporary chimpanzees, and they were vegetarians like the other apes—though no doubt supplementing this with the occasional monkey or antelope kill just as chimpanzees do today. The australopithecines were an extremely successful group, with more than a dozen species occupying habitats above 1,000 meters (3,300 feet) altitude in eastern and southern Africa between six and two million years ago.

Why the australopithecines adopted a bipedal stance has been a matter of debate ever since their first fossils were discovered in southern Africa in the 1920s. Australopithecines were certainly not as good at walking bipedally as we are, but they were capable of sustained bipedal locomotion in a way neither of the two living African great apes can manage. The most convincing reason they became bipedal seems to be thermoregulation. When traveling across open woodlands between the forests that provided them with their nighttime refuges and the riverside floodplains that provided them with rich food sources during the day, they faced significant radiant heat from the sun. Recent models of the benefits of bipedalism over quadrupedalism suggest that bipeds would have accumulated around 12% less heat compared to animals walking on all fours, mainly because only the top of their head and shoulders are exposed to the sun rather than the whole of the back. Even so, at this stage, they were confined to high-altitude habitats and would have been prevented by heat overload from coming down to coastal areas (just as chimpanzees and gorillas are obliged to remain at higher altitudes today). That seems to be why they never escaped from Africa: unable to cope with the heat in coastal areas, they had no way of accessing the land bridges over to Europe and Asia.

Paradoxically, the climate cooling that occurred around 2.5 million years ago seems to have caused them problems, possibly because their existing habitats were now much colder and they weren't able to cope with this. The australopithecines disappeared quite soon after this and were replaced by a new genus, *Homo* (the genus to which we belong), that evolved out of one of the australopithecine lineages at around this time.

This lineage had a skeleton that was even better adapted to bipedal travel, allowing a more nomadic lifestyle with the possibility of traveling considerable distances each day. It also had a significantly larger brain and began to develop simple tools manufactured from stone.

It was at this point that early humans seem to have lost their fur and gained the capacity to sweat as a way of increasing heat dissipation by evaporative cooling. This allowed them to colonize lower-lying coastal habitats, and so to escape Africa for the first time. Within a few tens of thousands of years, they had colonized most of southern Europe and Asia. Perhaps because they were more mobile and were able to maintain higher rates of gene flow between populations (see Question 43), they didn't speciate as much as the australopithecines had, and instead produced only two species, *Homo ergaster* in Africa and *Homo erectus* in Eurasia.

These early *Homo* species were extremely successful, surviving almost unchanged for around a million and a half years in Africa and nearly two million years in Asia. Around 500,000 years ago, however, some new lineages arose out of the African stock. Once again, there seems to have been a flurry of new species scattered in different parts of Africa, with even bigger brains and a more robust body than their predecessors. These gave rise to three very successful lineages: the Heidelberg folk in Africa and Europe, the Neanderthals in Europe and western Asia, and the Denisovan people in the Far East. These three lineages, collectively known as archaic humans, occupied most of Europe, Asia, the southern rim of the Mediterranean, and most of Africa for the better part of 500,000 years. They also produced more sophisticated tools than any of their predecessors.

62. Who were the Neanderthals?

Around 400,000 years ago, a distinctively new group of humans evolved in southern Europe from the archaic humans that had gradually replaced the ancestral *Homo* populations of Africa and Europe. They were heavily built and large-brained, with an elongated skull that terminated in a distinctive bun at the back. They were named the Neanderthals after the location in Germany where their fossils were first discovered in 1856. Neanderthals proved to be very successful, occupying southern Europe and the Near East for the better part of 350,000 years in the face of some very tough climatic conditions generated by successive Ice Ages. It was only with the last, and very deepest, of the Ice Ages that they seemed to falter.

The Neanderthals were stocky, with short arms and legs (at least compared to ours) and a barrel-shaped chest. As might be expected of a people living at high latitudes, they were light-skinned (see Question 68), and some of them at least had the gene for red hair. Their stocky build is thought to be related to their occupation of cold habitats: a more compact build with shorter limbs reduces heat loss—a phenomenon known as *Bergman's rule* and frequently seen in high-latitude mammals as well as contemporary Eskimo.

Their physique, however, gave them enormous strength—you certainly wouldn't have wanted a wrestling match with any of them. Perhaps because of this, their principal style of hunting involved surrounding a large prey animal and using heavy thrusting spears at close quarters. That this was a dangerous business is clear from the frequency of broken bones in Neanderthal skeletons. Theirs was a heavily meat-based diet, derived from hunting seriously large-bodied herbivores like rhinos and mammoths.

The Neanderthals' dependence on meat has been demonstrated by one of the many ingenious applications of basic physics and chemistry to archaeology. What you eat provides the raw material for building your bones, so its signature remains locked in your body ever afterward. Nitrogen is a particularly important feature of the diet because it is one of the core components of amino acids and hence proteins. Like many elements, it comes in two isotopes that differ slightly in atomic weight, with one being much more common than the other. Plants acquire nitrogen from the air or soil and then pass it on to the herbivores that feed on them, who in turn pass it on up the trophic scale (see Question 53) to the carnivores that feed on them. During this process, the heavier isotope is preserved more easily than the lighter one, and so accumulates disproportionately through successive trophic levels, allowing us to determine which trophic level a species belongs to. An analysis of nitrogen isotopes in west European Neanderthals from around 40,000 years ago has revealed that 80% of their diet was meat and just 20% was vegetable matter. This is similar to what we see in wolves (a top predator from the same time period) (see Question 52) and much higher than we find in the roughly 50% meat diet of modern hunter-gatherers.

Neanderthals developed much more sophisticated tools than their predecessors, although they never really aspired to anything like the kinds of sophisticated artwork that modern humans were beginning to produce from even as early as 100,000 years ago (see Question 66). Even though their hunting style depended on a group of men cooperating in a very dangerous hunt, they lived at relatively low population densities. Like all the archaics, they were polygamous, with women moving (or being moved) between social groups—as analysis of the genetics of some archaic populations reveal.

63. Why did the Neanderthals have such big eyes?

The most distinctive feature of the Neanderthals is the "bun" at the back of their head. Surprisingly, no suggestion has ever been made as to why they had this unusual feature. In fact, the bun is related to something else that is distinctive about Neanderthals—their unusually large eye sockets. Although their large eyes have occasionally attracted comment in the past, again no explanation has ever been offered for these. Both traits just seem to have been taken for granted. That's just how Neanderthals were.

However, the large eyes and the Neanderthal bun turn out to be linked. The explanation is directly related to something else that defines the Neanderthals—the fact that they lived at high latitudes north of the Tropic of Cancer (which currently lies at ~23° north of the equator). The problem with living outside the tropics is that days are often dull (sunlight has to pass through more atmosphere because it is coming in at an angle *and*, as we northerners know only too well, there is usually much more cloud cover). Moreover, day length becomes progressively more seasonal with increasing distance from the equator, with long dark winter nights. All this puts a premium on vision, and species that live at these latitudes tend to have better-than-average vision.

There is only one way to improve vision under these conditions, and that is to increase the area of the retina (the light-sensitive layer at the back of the eye) (see Question 12), so that more light is received. However, this means that the whole eyeball has to be bigger, and in turn this means that the neural mechanisms for processing the light signals received (both in the neurons that connect the back of the eyeball to the brain and the areas in the brain that process and interpret these signals) have to be proportionately larger—there is no point in having an efficient light-receiving mechanism and then not having the computer power to make sense of the extra signals. As a result, not only do you have bigger eyes but you need a bigger brain. Since the primary vision areas that process these signals are at the back of the brain, what you get is a Neanderthal bun.

In fact, we can detect this effect even in modern humans: populations that live at high latitudes have bigger eyeballs than people from the tropics (though nowhere near as large as those of the Neanderthals), and they have a larger brain as a result. (We don't have a bun as such, because modern humans haven't evolved anything like as large a visual system as the Neanderthals did.) This latitudinal difference in modern human brain volumes has been known about for a very long time, though its significance has invariably been misunderstood. It has nothing to do with the intelligence of tropical versus high-latitude peoples, since intelligence is associated with the frontal lobes and these do not differ in size. It simply has to do with visual acuity. In fact, in modern humans, the size of the eyeball (and hence the brain) correlates well with ambient light levels (how dull it is on the average day) under which a population lives. About the only thing it does mean is that people from the tropics cannot see as well on dull days when they go to live at high latitudes, and people from high latitudes have their eyeballs burned out when they go to the tropics (which is why they are much more likely to wear sunglasses).

Modern humans have lived in Europe for only around 40,000 years—only a tenth of the time the Neanderthals were there. It is quite remarkable that we can pick up such a clear signal of this adaptation after so short a period of time. It reflects the speed at which selection can drive evolution when the demands of the environment are high enough (see Question 16). It is also a reminder that many of these effects are driven by climate (see Question 17).

Despite occasional claims that Neanderthals had bigger brains than we do, in fact their average brain size was much the same as ours, perhaps even a little smaller. One consequence of this is that by devoting so much brain space to visual processing, they would have had less neural matter in the forebrain—where most of our smart thinking is done. This is probably why they had a much more sloped forehead than modern humans do with our high-domed head designed to accommodate our large forebrain. This may explain why, despite having a brain as big as ours, the stone tools they made never developed to quite the same level of sophistication as those of the anatomically modern humans who replaced them (the increasingly frequent claims that Neanderthals were fully modern in all senses notwithstanding). This is not, of course, to say that the Neanderthals were stupid. Far from it. They would never have been able to survive the rigors of the Ice

Ages in Europe for almost 400,000 years if they had not been smart and able to figure out ways to cope with these kinds of conditions.

What it does mean, however, is that, while a lot smarter than their predecessors, Neanderthals were not as smart as the anatomically modern humans that evolved in Africa around 150,000 years ago and arrived in the Neanderthal heartlands of Europe around 40,000 years ago. These Africans arrived with their much darker skin (and in all likelihood curly hair) (see Question 68), their more sophisticated miniaturized tools like needles and buttons, novel weapons like bows and arrows, and javelin-like spears with spear-throwers that greatly increased the distance and accuracy that a spear could be thrown. The smaller frontal lobes of the Neanderthal brain would also have meant that they were socially less smart, and would not have been able to maintain such large communities as modern humans (see Question 64). And that would have had all kinds of ramifications impinging directly on their ability to survive.

64. Why did the Neanderthals go extinct?

Having occupied southern Europe as far as the western edge of Siberia for most of the last 400,000 years, the last of the Neanderthals died out in Spain shortly after 40,000 years ago—quite soon after modern humans entered Europe from the Russian steppes. It had been thought that they had survived until as late as 28,000 years ago in Spain, and hence overlapped with modern humans for the better part of 15,000 years, but recent revised datings for their last known occupation sites suggest that these are older than originally thought. In fact, by the time modern humans arrived from the Russian steppes, the Neanderthals were already probably so thin on the ground that the modern humans effectively came into a landscape that was empty of competitors. After all, for well over 100,000 years the Neanderthals in the Levant had blocked modern humans' ability to exit Africa via the only land bridge, forcing them to use the much more difficult Bab el-Mandeb sea crossing at the bottom end of the Arabian peninsula. So it seems unlikely that modern humans could have so easily invaded Europe had it been densely populated by physically stronger Neanderthals.

Quite why the Neanderthals died out has been the subject of intense debate for over a century. Among the many explanations that have been offered have been extermination by modern humans, diseases brought in by modern humans to which they had no immunity, inability to cope as well as modern humans with the deteriorating climate at the height of the last Ice Age, population sizes that were too small to replace themselves when local communities went extinct, being absorbed into modern human populations through interbreeding, and being overwhelmed by a volcanic catastrophe. All of these may be true—at least to some extent—though we still need to ask whether one was more important than the others. Nonetheless, they beg the question as to why modern humans survived in the same habitats that put an end to the Neanderthals.

Direct extermination of Neanderthals by modern humans seems unlikely, given that Neanderthals were physically much stronger than modern humans. Nonetheless, it is surely no coincidence that Neanderthals died out within a few thousand years of modern humans' arrival. The collapse of the American Indian populations in the sixteenth and seventeenth centuries caused by European diseases to which they had no immunity perhaps provides a clue to what might have happened. The same historical examples suggest that this might have been helped by a little active extermination here and there and, no doubt, by the occasional stealing of women (see Question 65).

If novel diseases resulted in a further collapse of the population, this would have pushed the Neanderthals into small isolated communities, even in the absence of competition from modern humans. Population fragmentation of this kind is perhaps the single most important immediate threat to any species' survival since its populations then lack the resilience to recover from other demographic threats (see Question 45). Any slight extra ecological pressure from modern humans, any slight reduction in the availability of their normal prey species, any worsening of the climate, a few less women to reproduce—any one of these could have tipped them over the brink in contexts they might easily have survived had their populations not been so small and so isolated.

One indicator that Neanderthals would have had smaller populations than modern humans comes from the size of their brain. The social brain hypothesis (see Question 85) defines a relationship between community size and the volume of the frontal lobes (in particular) in primates (including modern humans). Because more of the Neanderthal brain was devoted to the occipital lobes and visual processing and much less to frontal lobes and smart/social cognition, their smaller frontal lobes would have supported communities of around 120 individuals compared to the 150 characteristic of anatomically modern humans. In any conflict that might have occurred, modern humans would have been able to draw on a larger numbers of fighters. It would also have meant that they were just not quite in the same cultural league as modern humans. This would have resulted in a significant difference in the way culture was used to bond social communities (see Question 87) and hence the size of these communities.

Clothing might have played a particularly important role in their ability to cope with their deteriorating environment. Avoiding heat loss is crucial for survival in cold climates, and the differences between seamed and buttoned clothing as opposed to simple wrap-around capes might have made all the difference between modern humans and Neanderthals if these reflected differences in technical imagination. A difference in the thermal properties of winter clothing may explain why modern humans were able to live much farther north than Neanderthals during the Ice Ages, irrespective of the way both species were pushed southward and northward by the advancing and retreating ice fronts. It may have been key to the one species' ability to survive and the other's demise.

There is archeological evidence for buttons and beading attached to garments from sites in Europe occupied by modern humans dated to 35,000 years ago (and earlier than that in Africa), indicating that modern humans must have had the needles and awls needed for making more sophisticated clothing well before the Neanderthals disappeared. There is no evidence that Neanderthals ever had these kinds of tools, even though their tools were getting somewhat more sophisticated toward the end.

Some unexpected light on this question is provided by the genetics of the two louse species that infest modern humans (body lice and head lice, which, incidentally, do not interbreed because they occupy different parts of the body).[1] The genetics of the two louse species indicates that they had a common origin as a single species around 100,000 years ago. Lice can survive only in enclosed environments (usually provided by fur). Our hairless bodies were a louse desert from two million years ago when we lost our body hair (see Question 61), with our hairy head being the only remaining haven for them. Once we had invented close-fitting clothing, however, a new habitat magically became available and allowed some of them to migrate to new empty habitat. Once there, they inevitably underwent their own evolution and became a separate species. The time at which they became two species gives us an estimate of the likely time at which we invented clothing.

65. Is it true that we all have Neanderthal genes?

The Neanderthals have had a checkered history, with opinion swinging back and forth between the suggestion that they were our direct ancestors (and therefore obviously smart) and the suggestion that they were a side branch that went extinct (the dumb cavemen). Advances in molecular genetics in the last decade, however, have suggested that they are *not* our direct ancestors. Nonetheless, around 1–4% of the genes of modern Europeans *are* derived from Neanderthals, presumably by intermarriage (or more likely, given historical human behavior, Neanderthal women captured by human bands). There are no Neanderthal genes at all in any of the African genetic lineages because there were no Neanderthals in Africa. Any interbreeding with modern humans happened after modern humans had left Africa.

The genes that Europeans inherited from Neanderthals are ones that influence traits such as lighter hair and skin color, a tendency to tan and burn, such behaviors as being a night owl, and a predilection for nicotine. (This is not evidence that either Neanderthals or modern humans smoked tobacco 40,000 years ago: the tobacco plant is native to the Americas, and humans didn't get there until 25,000 years later.) There is also evidence to suggest Neanderthal genes influence such traits as proneness to daytime napping, loneliness, and low mood. Since skin color is related to latitude (lighter skin shades facilitate the production of vitamin D in the skin by sunlight, which is essential for survival at high latitudes) (see Question 68), the Neanderthals, having spent several hundred thousand years in Europe, were lighter-skinned. In contrast, the incoming modern humans would have had the darker skin color typical of peoples from Africa. Acquiring the Neanderthals' genes for light skin color would have greatly speeded up their adaptation to life at high latitudes (I'll explain why in Question 68).

More surprising still, perhaps, has been the subsequent discovery that east Asians and modern Melanesians from New Guinea and associated islands, as well as the Negrito populations from the Philippines, have 4–8% of their genes from another group of archaic humans, the Denisovans. The Denisovans are known only from a few teeth found in a cave in the Altai Mountains in central Siberia representing four different individuals who lived around 40,000 years ago. The Denisovans appear to have been an even earlier offshoot from the archaic Heidelberg populations of Europe that had moved much farther eastward. Modern human populations invading southern Asia some 70,000 years ago must have met up with them in southeast Asia and interbred with them. In fact, the genetic evidence suggests that this may have involved at least two separate waves of gene exchange.

66. Why did only one species from our entire lineage survive?

Because we are the only species from our lineage that is currently alive, we are apt to assume it has always been this way, with one species being replaced by its descendant in a series of temporal speciation events (see Question 43). In fact, the last 40,000 years since the Neanderthals disappeared have been very unusual because there has been only one species of our lineage around. The apparent contrast between us and our nearest living relatives, the chimpanzees, has been exaggerated by the absence of any intermediates.

Between two and four million years ago when the australopithecines were at their height, there were as many as four species living at the same time. Between 1.8 and 2.4 million years ago, around the time when the genus *Homo* first evolved, there were several periods when up to six different species of our lineage were alive at the same time. In some cases, early *Homo* species coexisted in the same habitats with late australopithecines. The next million years saw this reduced to just two species. But, from around half a million years ago, there were as many as six hominin species alive in different parts of Africa and Eurasia.[2] Then, from around 50,000 years ago, this dropped rapidly to two and then just us.

For the first 100,000 years of its existence as a distinct species, our species was confined to Africa, most likely because the Neanderthals in the Near East blocked their route out of Africa. Around 70,000 years ago, however, small numbers of individuals were able to cross over into southern Asia (probably using boats to cross the Bab el-Mandeb at the southern end of the Red Sea beyond the reach of the Neanderthals).

From there, they spread very rapidly across southern Asia as far as Australia over the next 20,000 years, almost certainly by coastal hopping (possibly in boats). From southern Asia, they looped back into Europe around 40,000 years ago through modern-day Russia and, after briefly coexisting with the Neanderthals, rapidly replaced this older species to become the only human species on the planet.

Quite why we have ended up as the only species of our lineage remains an unresolved puzzle, although there can be little doubt that the climatic conditions of the last Ice Age were imposing significant stresses on all the human species still around at the time. The last of the *Homo erectus* species in southern China and the diminutive one-meter (three-foot) tall Hobbit (*Homo floresiensis*) of the Indonesian islands all died out around 60,000 years ago when the last Ice Age was at its height (although, in both cases, suspiciously close to the time at which modern humans arrived from Africa). Despite using caves as refuges and having had control over fire from around 400,000 years ago, it seems that the Neanderthals were unable to cope with the cold as well as anatomically modern humans (see Question 64).

One thing that is clear, however, is that the birth of our species was finely balanced and we could easily have disappeared along with all the other hominin species. This is clear from the genetic evidence for the dramatic population bottleneck that gave rise to modern humans: statistical analysis of the distribution of contemporary human genes tells us that everyone alive today is descended from a set of just 5,000 breeding females who lived around 150,000 years ago (see Question 30). In many ways, it is a miracle that any of us are here at all.

The history encapsulated in the genes of living humans tells us that one of the four major mitochondrial DNA (mtDNA) lineages present in Africa (the one with its origins in eastern Africa that gave rise to Europeans, Asians, and Native Americans, as well as the later historical expansion of the Bantu peoples into western and southern Africa) underwent a dramatic expansion in population size about 10,000 years before it left Africa to populate the rest of the habitable world around 70,000 years ago. This lineage now accounts for half of all the people in sub-Saharan Africa as well as all the native inhabitants of the rest of the world. The other three mtDNA lineages remained within Africa and did not exhibit any real expansion until much later, if at all.

67. What has molecular genetics been able to tell us about our recent history?

Surprising as it may seem, molecular genetics can tell us a great deal about the recent history of our species. We have already seen that our genetics tells us that we are a relatively young species with its origins in a very small group of people (see Question 66). As a result, the genetic variability across living humans is really quite modest compared to most species. We just haven't had enough time to diversify. A second finding is that all modern Europeans (Caucasians), Australians, Asians, and Native Americans are more closely related to each other than they are to Africans in general. Non-Europeans have much less genetic variation than the native populations of Africa. These non-African populations have a common ancestor around 70,000 years ago, probably from the northeast corner of Africa, whereas the common ancestor of all Africans lived around 150,000 years ago.

Genetic analysis tells us that the Inuit (Eskimo) and Amerindian populations represent a subset of the Eurasian population that slipped across the Bering Strait between eastern Siberia and Alaska as recently as 16,000 years ago when the sea levels were much lower and there was a land bridge between the two continents. In fact, the linguistic evidence suggests that there were three major waves, yielding three major groups of related languages: an early wave that gradually spread down into present-day United States and eventually South America, yielding a large, diffuse family of Amerindian languages; a second wave that gave rise to the Na-Dene language group, which includes the Crow and other tribes from southern Canada and the Apache and Navaho of the southwestern United States; and finally the Inuit, or Eskimo, who are the most recent arrivals and are confined to the upper parts of Canada and Greenland (which they reached only around 2,000 years ago).

Molecular genetics has also been able to tell us many surprising things about recent history. An analysis of Y chromosomes (inherited only via the male line) from males from different parts of the world suggests that half a percent of all the males alive today are the direct descendants of the thirteenth-century Mongol leader Genghis Khan and his brothers, with a figure closer to 7% within the territorial extent of the original Mongol Empire in Asia and eastern Europe. This is a direct consequence of the historically well documented fact that when the Mongols captured a city, they invariably slaughtered all the men and enslaved all the women. Genghis Khan and his lieutenants were the principal beneficiaries.

Similarly, an analysis of the Y chromosomes of living Icelanders reveals a classic Norse origin, but 85% of Icelandic women's mitochondrial DNA (inherited only through the maternal line) has a Celtic origin. Evidently, when the Norse settled Iceland in the ninth and tenth centuries, their own women were less than enthusiastic about the idea of accompanying them, so the men helped themselves to some (probably less-than-willing) women from Ireland and Scotland on the way. Indeed, the Icelandic sagas (family histories, most of which were written down in the twelfth century) repeatedly mention "Irish" female slaves—although that could well just mean they acquired them in Dublin, the capital of the western Viking kingdom and the main slave market for Celts and Saxons captured during Viking raids on the British mainland.[3]

Another example is provided by a comparison of contemporary Y chromosome and mitochondrial DNA on a transect across southern England from East Anglia in the east to Wales in the west. While the women's mtDNA indicates a fairly consistent predominantly Celtic origin, the men's Y chromosome DNA reveals a cline with mainly Saxon genes in the east (the point of entry of the Anglo-Saxon invaders of the fifth century AD) to mainly Celtic genes in the west (where the Anglo-Saxons didn't penetrate). In other words, as they spread slowly across southern England, the Anglo-Saxons disposed of the Celtic men and stole their women.

Another historical example is provided by the Phoenicians. This Semitic civilization existed in modern Lebanon between 1500 and 300 BC and created one of the greatest trading empires in the Mediterranean world. They established colonies in coastal ports all over the Mediterranean, from Turkey to Spain. Genetic studies suggest that a Y chromosome genetic signature that is common today in Lebanon (the Phoenician homeland) can be identified in many of the major trading ports around the Mediterranean that are known from historical sources to have been associated with the Phoenicians.

68. Are racial differences adaptive?

The term "race" has been used quite loosely in biology. In the nineteenth century, it could refer to a species or to some subdivision of a species. It tends to be used now to refer to local populations of a species or subspecies when these differ sufficiently in appearance or genetic traits to mark them as distinctive. Some anatomical traits can adapt quite quickly to local environmental conditions or through sexual selection. In other cases, genetic drift may lead to noticeable differences if populations that live at opposite ends of a species' geographical distribution have been separated for long enough. Its common use in humans to refer to major continental groupings (Africans, Europeans, Asians, etc.) is technically incorrect.

Africans are not a single race, but rather many different races: they are made up of four major mtDNA subdivisions (or *haplogroups*) that originated in different parts of Africa, one of which includes everyone else from outside Africa. If one really wanted to do so, one could build a biologically plausible case for there being four human races (the four African mtDNA haplogroups). One of these would also contain all the European and circum-Mediterranean peoples, Asians, native Australians, native Americans, the Cushitic-speaking pastoralist tribes of eastern Africa (such as the Ethiopian Amhara, the Maasai, El Molo, and Iraqw), the Fulani-speaking peoples of the Sahel (mostly pastoralists, who appear to be historical immigrants from north Africa), and the Bantu (the largest single group of African tribes, who spread over much of central and southern Africa from their homeland in the region of Cameroon during a great expansion beginning some 3,000–4,000 years ago). In short, the biologically meaningful races do not bear much resemblance to the way the term has been commonly used since the nineteenth century.

Whether or not human races exist, we can still ask whether any of the differences between the various human populations are adaptive. Skin color, for example, is an adaptation to the local levels of ultraviolet radiation in sunlight, and has little to do with racial groupings as such. It is an adaptation to the risk of cosmic ray damage to the skin and internal organs, this being a function of latitude and altitude. High densities of melanocytes (the cells in the skin that produce the melanin that gives skin its dark color) protect the skin from excess sunlight. Hence, populations living in the strong sunlight regions of the tropics are characteristically dark-skinned, while populations that live farther from the tropics have lighter skin color because the sunlight they experience is weaker. By the same token, populations that live at high altitudes, in Tibet or the Andes for example, have darker skin than you would expect for their latitudes because, being that much closer to the sun with a very thin atmosphere above them, they are subject to more intense radiation from sunlight.

Prior to modern humans' emergence out of Africa some 70,000 years ago, all our ancestors would have been dark-skinned (though probably not as dark as some modern Bantu populations). Outside the tropics, however, dark skin becomes a liability because, under the poor light conditions that prevail at high latitudes, not enough sunlight gets through the skin to allow it to synthesize vitamin D, which is essential for the intestinal absorption of calcium and other minerals. Indeed, in darker-skinned tropical and subtropical populations, women (and babies) are always much lighter-skinned than their menfolk because they need to be able to maximize the uptake of calcium for the baby's growing bones. The intensity of selection is sufficiently strong that populations that migrate to high latitudes from the tropics undergo significant loss of melanocytes in about 2,500 years, or about 100 generations. Some Amazonian Indians are lighter-skinned than would be expected for the latitude at which they live, and it has been suggested that this is because they have only relatively recently moved into the area.

Another example of a local adaptation includes body shape. Peoples who live in open savannah regions in the tropics where they are exposed to direct sunlight tend to be tall and thin so as to minimize the skin surface exposed to sunlight at the hottest times of the day when the sun is overhead. Examples include many of the tropical pastoralist peoples like the East African Maasai and the Nuer of the Sudan. In contrast, people who live in the deep shade of tropical forests like the pygmies of Central Africa and the Negrito peoples of South Asia tend to be short and lightly built.

Other adaptations include the lactase enzyme that converts the lactose sugars in milk into more accessible glucoses (see Question 16). Only Caucasians (Indo-Europeans) and the cattle-keeping African pastoralists have the mutation that allows them to digest milk as adults. Other races can consume milk products only if they are processed first (e.g., as yogurt or cheese, where the bacteria responsible for fermentation convert the lactose into more digestible lactic acid).

Another example of a genetic adaptation is the sickle cell gene from West Africa and the closely related thalassemia gene in the eastern Mediterranean, both of which provide resistance to the malaria parasite in the heterozygous form (a sickle cell gene combined with a normal gene)—albeit at the expense of an excruciatingly painful and completely debilitating condition in the homozygous recessive form (where both genes are of the sickle type) (see Question 22) that leads to early death. These mutations are found only in regions where malaria-bearing mosquitoes are found, mainly because the selection against the sickle cell mutant is otherwise so intense.

69. Why are humans the only species with language?

Although they are often lumped together, it's important to distinguish between language and speech. Language is a cognitive mechanism whereby we assign names to particular mental states—sometimes known as the "language of thought." Mammals (and probably birds) naturally store information about the world in the form of propositions (*X causes Y*) rather than simple facts. When language has grammatical structure (as it does in human languages), that allows us to formulate complex (multi-proposition) sentences that both help us organize our thoughts and allow us to make inferences (*Some X are Y, and some Y are Z; therefore some X are Z*). Speech is the physical mechanism we use to transmit those thoughts to other individuals. Speech is mainly a vocal medium, though alternatives such as sign language are possible (albeit rarely anything like as rich). We might well not be the only species that can formulate thoughts in our minds, but we are the only species that has speech (as opposed to being able to imitate speech sounds in the way that parrots can do).

Speech is all about breath control (the ability to breathe out in a controlled way over a period of a minute or more). Without that, speech as we know it would be impossible since we could otherwise produce only very simple sentences. This did not

appear as a single grand mutation. Rather, it was the culmi-
nation of a very long series of accidental evolutionary steps
unrelated to each other that began when our ape ancestors first
began to walk bipedally some six or seven million years ago.
Walking on all fours locks the rib cage (to support the body)
whenever an arm is in contact with the ground. This makes it
impossible for monkeys (or any other species that walks on all
fours) to breathe while they walk. In effect, they have to alternate each step with a breath. Speech is possible only when the
pressure is removed from the rib cage and breathing is independent of standing and locomotion.

The second stage occurred when archaic humans evolved
the capacity to sing, most likely in the form of wordless
humming as a mechanism for social bonding (see Question
96). The anatomical changes associated with full breath control are indicated by the appearance of a significantly enlarged
neural tract in the upper chest (where the nerves that control
the diaphragm and chest wall muscles emerge from the spine)
and a shift in the position of the hyoid bone in the throat (the
delicate bone that supports the top of the esophagus, which
moved down so as to increase the size of the vocal chamber
in the mouth and throat). We share these traits with the
Neanderthals, so they certainly date back at least 500,000 years
and are likely to have been shared by all archaic humans.

The next two phases were the linking of the breathing appa-
ratus with the thoughts in the mind, allowing thoughts to be
expressed in speech as words and then, importantly, the ability
to construct embedded propositions (in effect grammar).
These are two quite separate processes. Since both happen in
the brain, they leave no fossil record. However, we know from
child development that they emerge in this order. Children
start with simple sentences, often just instructions ("Come!,"
"More!") that are about as complicated as the communications of monkeys and apes. They then gradually develop more
and more complex sentences. The development and evolution
of complex grammatically structured sentences with several

clauses (or propositions) depend on a cognitive process known as *mindreading*, or mentalizing.

Mentalizing is the ability to understand what someone else is thinking. Children are not born with the capacity to mentalize, but they develop it from about age five. Prior to that, they know their own minds but cannot distinguish the content of their minds from someone else's. From age five, they appreciate that others might believe something different from what they believe ("I believe that Jim thinks the world is flat [even though I know it is round]"). After that, they develop the capacity to progressively include additional mind states until they reach the normal adult limit at about five mind states sometime during the teenage years. The more mind states we can handle, the more grammatically complex the sentences we can unpack.

The fact that mentalizing abilities correlate with frontal lobe volume (even in normal adults) means that we can use this relationship to provide a glimpse back into the fossil record. Doing so suggests that archaic humans (including Neanderthals) would have been able to achieve only four levels of mentalizing (compared to five in modern humans). So although Neanderthals are very likely to have been able to speak, their language would not have been as complicated as ours. This means their jokes would have been less funny, their stories less complex, and their ability to figure out complicated sequences of causal relationships poorer. Fully modern language as we know it did not, it seems, emerge until the appearance of anatomically modern humans around 150,000 years ago.

70. Are humans still evolving?

So long as individuals vary in the rates at which they produce grandchildren, there will be evolution. And so long as the environment changes over time so as to favor some individuals, or some species, over others, there will be evolution. The rate of evolution can, of course, slow down, but it can never cease altogether. Modern humans have been able to slow down the rate of evolution to some extent through improvements in hygiene and medicine that help to preserve life, thereby allowing individuals who might previously not have been able to reproduce or rear offspring to do so. However, this assumes that there will be no major crises such as major wars, famines, mass exterminations, or other catastrophes that will cause differential mortality on a very large scale.

A second factor that has slowed the rate of evolution in modern humans is the dramatic improvements in transport over the last few centuries. Since geographical isolation is a major factor in speciation (see Question 43), the dramatic spread of humans around the world would, given long enough, eventually have led to different geographical populations evolving into separate species (though it would probably have taken several hundred thousand years more of isolation). However, gene flow between different human populations has increased dramatically in the last few thousand years as a result of greatly improved sea, land, and air transport, and the resulting opportunities this has brought for intermarriage. Even so, there are some surprising examples in which quite modest geographical or even social barriers have prevented the intermixing of neighboring populations.

One example is provided by the River Tamar that separates the counties of Devon and Cornwall in the southwest of England. Culturally, the river has always separated two populations who view each other with suspicion (and still do). Amazingly, the River Tamar turns out to demarcate two surprisingly well-differentiated genetic populations. On one side are the descendants of the original Celtic inhabitants of Britain (who, until it became extinct in the eighteenth century, spoke a Celtic language, Cornish) and on the other (Devon) side the descendants of sixth-century Anglo-Saxon invaders. (For more on what molecular genetics has told us about recent history, see Question 67.) Despite the fact that the original circumstances of the divide have long ceased to have any relevance, these two neighboring populations have maintained sufficient local reproductive isolation for a genetic difference still to be present after fourteen centuries.

Another example is the Hindu caste system in India. It has long been known that the four main castes, or *varnas*, derive from the social system of the Indo-European invaders who arrived in northern India from the Russian steps some four thousand years ago, with the lowest class of *dalits* (the Untouchables) representing the original inhabitants they conquered. Although the castes were characterized by high levels of endogamy (marrying within caste), the system has latterly been viewed as being largely social. In fact, it turns out that the proportion of Indo-European genes (in effect, those shared with other Europeans) correlates with caste status: the highest caste, the Brahmin or priestly caste, which has assiduously maintained its identity by rigorous practices of purity, has the highest proportion of Indo-European genes. The *dalits*, as the original inhabitants, naturally have the fewest. Remarkably, these genetic differences have been maintained by cultural mechanisms across 4,000 years with only very modest gene flow—despite the frequency with which northern India has been invaded and conquered by Persians, Afghans, Arabs, and Mongols.

One interesting question is whether our increasing dependence on sophisticated knowledge and complex technology might lead to the evolution of larger brains in our species. The answer is: probably not, and for two very good reasons. One is that our brain size is already at the absolute limit for natural births. In mammals as a whole, babies are born when they are more or less capable of fending for themselves and this is mainly determined by the baby's brain reaching a certain level of development. By this token, humans should in fact have a pregnancy of 21 months, about the same as the elephant with its similar-sized brain.

When we adopted a bipedal stance some six million years ago, the resulting change in the shape of the pelvis (required to support the trunk) dramatically reduced the size of the pelvic inlet (or birth canal). This wasn't a big problem until, five and a half million years later, we decided that larger brains were a good thing. We then faced the problem of squeezing a very large head through a very small hole. Our solution to this small difficulty was to give birth to very premature babies and allow them to complete brain growth outside the womb. Human babies do not reach the same level of development as newborn ape babies until they are about a year old (the equivalent of 21 months gestation). Shortening the length of pregnancy any further isn't really an option, because human babies are already on the very edge of survival when they are born.

Nothing being impossible in evolution, there is, of course, an alternative solution, and that is to increase the size of the female pelvis so as to make the birth canal larger. However, a wider pelvis would significantly impact women's walking and running, causing them to waddle rather than walk. Aside from reducing the likelihood of any more women's athletic records ever being broken, it would make women considerably less mobile. If nothing else, that would likely lead to increased segregation between the sexes—as essentially happens in some Sahelian cultures where the girls are force-fed to make them overweight and ostensibly more attractive, or in historical China where foot-binding confined upper-class women to the home. I suspect it is not a solution that would go down especially well.

8
EVOLUTION OF BEHAVIOR

71. What role does behavior play in evolution?

Most adaptations ultimately require some level of genetic change, but genetic change depends in part on the availability of sufficient genetic variation within the population—or, at the very least, the chance to produce new mutations. If the environmental change is fast enough—like the Younger Dryas climatic event that marked the end of the last Ice Age (see Question 17)—a species can easily go extinct, especially if it has a long generation time and breeds slowly (see Question 45). When the change is slower, however, being able to adapt behaviorally buys time in a way that can allow a species to avoid extinction long enough to evolve an appropriate genetic response. This is known as the Baldwin effect, after the American psychologist James Mark Baldwin who proposed it in 1896.[1]

The ability to recognize that there are regularities in the environment and act on these is especially important when environments undergo constant change, as is especially true of terrestrial habitats. Being able to learn about cues in the environment, such as knowing where the safe sleeping sites are or knowing which foods are edible and which not, can make the difference between life and death. Optimal foraging theory (see Question 12) is only possible if animals can learn about environmental cues and make sensible decisions about what to eat or where to go based on these.

Another context in which strategic decision making becomes especially important is mating (see Questions 56 and 74). When a species consists of two subgroups (i.e., sexes) that have different reproductive interests, there is scope for manipulation and strategizing, and this will create an arms race as the two subgroups attempt to manipulate and outwit each other in order to gain an evolutionary advantage. Indeed, any social situation undergoes constant change over time, and animals need to be able to track these changes and make inferences about the likely future behavior of other individuals.

Behavioral adaptation requires the evolution of a brain that is complex enough to make the kinds of decisions needed to allow an animal to adjust its activities to changing conditions. In addition, of course, the animal needs a perceptual system complex enough to provide the brain with the inputs necessary for these calculations. In other words, the point of having a big brain is to be behaviorally flexible so that you can adapt your behavior at least to the small-scale changes in the environment.

Brains capable of this do not come for free, however. Neurons are extremely expensive to evolve and even more expensive to maintain in readiness to fire—at rest, the brain consumes around 10 times more energy than muscle does. This is largely due to the nature of the electrochemical mechanism that enables a neuron to fire when activated by input from another nerve. There are two separate components to this mechanism: (1) the energy cost of the sodium pump that maintains the electrical differential across the neural membrane that is necessary for it to fire and (2) the cost of replacing the chemically expensive neurotransmitters after the neuron has fired. But if you can find a solution to this constraint, then growing a big brain and using it to find better solutions to the problems of survival and successful reproduction has distinct advantages.

In short, behavioral flexibility is the key to evolutionary success, but it comes at a price.

72. How can we explain the evolution of altruism?

Darwin was broadly satisfied with his explanation of how species evolved but was troubled by the fact that he could not find a satisfactory explanation for the evolution of altruism (the sacrifice that one individual makes for the benefit of another). The pressure exerted by natural selection in promoting the traits of individuals that reproduce most successfully ought to result in traits like altruism being rapidly eliminated from the population. Yet, as Darwin knew from close observation of his own beehives, female honeybees are sterile workers that have given up all prospect of reproduction and, instead, labor assiduously for the reproductive benefit of their sister, the queen, who is the only female in the hive to reproduce.

The problem was not resolved until the 1960s, when Bill Hamilton had the inspired insight to realize that if fitness is about the future prospects of specific genes, then an individual could contribute copies of that gene to the next generation by helping a close relative reproduce—providing, of course, they had both inherited a copy of the gene from a common ancestor (his concept of inclusive fitness) (see Question 25). What he realized is that if the number of extra offspring produced by a relative as a direct result of your altruism, when devalued by the relatedness between you (i.e., the probability of sharing a specific gene), was greater than the number of future offspring you lost as a result of that action, then natural selection would favor the gene that promotes altruism. Moreover, you should sacrifice more for a close relative than for a distant one. This principle became known as the *theory of kin selection* because it argued that individuals should prioritize kin. Everything depends, however, on the ratio of the gains to losses *and* the degree of relatedness.

Unfortunately, discussions about altruism in humans invariably fall afoul of a peculiar form of linguistic slippage. In everyday life, altruism can range from small donations to panhandlers on the street (generous, but negligible fitness cost to the giver) to lending money to help someone achieve their aim in life (a bit more serious, but likely worth it for a relative) to giving up your life to save another (much as, in Dickens's novel *A Tale of Two Cities*, Sydney Carton went to the guillotine in revolutionary Paris in place of the French aristocrat Charles Darnay). The problem is that we invariably tend to think in terms of the last case when discussing altruism and assume that it represents the norm of behavior. It is then easy to conclude that humans behave in ways that are anti-Darwinian. We need to be careful with these kinds of claims for several reasons.

First, the size of the donation is crucial in any evolutionary explanation. The question we have to ask first is: who gives how much to whom how often? Loose change given to a beggar is hardly likely to tax most of us, let alone impact our reproductive prospects. We might choose to count it as everyday altruism (and, indeed, we often do), but it does not actually pass the test for biological altruism: the cost is simply too low. Giving away large amounts of money or resources may be a more substantive issue, but a large donation by someone who is very rich might not have that much impact on their fitness. We also need to look closely at the objectives involved. A donation by a male to an unrelated female of breeding age might be better explained as mate advertising: an everyday example would be taking someone out for an expensive meal in the hopes of impressing them sufficiently to yield a favorable outcome.

But what about those situations, like Sydney Carton's, in which the altruist actually sacrifices their life so that another might live? There are two questions we need to ask. First, how often does this actually happen in real life? The answer is: not very often, in fact—which is precisely why we make such a fuss on the rare occasions when it does happen. Dickens's story would not have been *so* poignant had Carton merely given a few French francs to defray the costs of Darnay's funeral—or even agreed to look after the man's niece after he had been guillotined. An analysis of the contexts in which civilians carried out dangerous rescues indicated rather starkly that men tended to put their lives at risk to rescue women of reproductive age and women tended to do so only when very close kin (such as their children) were involved—two perfectly sensible evolutionary reasons associated, respectively, with mating opportunities and parental investment.

This does not, of course, deny that we should applaud risk-taking when it happens. But we should not over-interpret its significance. Our actions may not always be what they seem at first sight, and we need to be careful how we interpret them.

73. Why has cooperation evolved?

Cooperation is, in many ways, the key to complex social life, and hence is of central importance in the behavior of humans and other socially advanced animals. There are many cases in nature where animals appear to cooperate (e.g., chimpanzees hunting monkeys, beavers building their dams), and it obviously occurs in humans. But from an evolutionary point of view, cooperation risks the same dilemma as altruism (see Question 72): if you cooperate with someone, you necessarily invest your time, effort, or money in their reproduction, and if they don't pay you back, then you have behaved altruistically. In economics, this problem is known as the *public goods dilemma*.

So how can cooperation evolve?

One suggestion, of course, is kin selection: even if you never repay me the favor I do for you, providing we are related my payback comes in the next generation through the extra offspring that you produce and my share in those (see Question

72). But that can apply only if we are relatives. If we are un-related, could cooperation still arise? One possibility is *mutualism* (or *group augmentation selection*): if the task requires us both to cooperate at the same time and we both benefit equally by doing so, then cooperation can easily evolve. One obvious example is living in groups as a defense against predation—where the mere presence of the group deters a predator from trying to attack and the animals do not themselves actively attack the predator (see Question 81).

Cooperative hunting is another example: by cooperating, we benefit by being able to bring down much larger prey than either of us could do on our own, as is the case for many large predators such as lions, hyenas, and wolves. However, this doesn't entirely remove the public goods dilemma problem: I might rush around looking extremely busy while avoiding taking any risks but still claim my share of the kill. Known as *freeriding*, this is a common problem, not just in nature but in everyday human social life. It's the person who borrows but never pays back, who accepts a social invitation but never reciprocates.

The conventional solution to this problem is punishment: if someone is not fulfilling their part of the social contract, we can punish them by applying the "win-stay/lose-shift" strategy: I'll continue to cooperate with you so long as you reciprocate, but the moment you fail to do so I won't cooperate again. This is possible, of course, only if the species has evolved a brain large enough to allow it to remember past encounters and their outcomes. An alternative version of this is for me to check your reputation before I help you out. That way, I can assess how reliable you are on the basis of how you have behaved toward others in the past. An experiment by the Canadian psychologist Yvan Russell showed that chimpanzees do monitor others' interactions with third parties and use that information to decide whether or not to cooperate. Another possibility is altruistic punishment, whereby an individual is willing to punish someone who fails to live up to the deal. In experiments with humans, people are often prepared to pay a fee to punish those who fail to behave well.

In most such cases, there will be some mechanism that allows the partners to monitor each other's adherence to the social contract. Humans are especially sensitive to people who renege on social arrangements and fair share arrangements. Even some animals do this. One example is the hedge sparrow, or dunnock, a small European garden bird. Because in birds the females are able to store sperm from matings with different males and then selectively fertilize their eggs with all or some of this stored sperm, clutches can often be fertilized by several different males, including (or not) the female's mate. The British ornithologist Nick Davies found that the male adjusts how much effort he devotes to foraging for grubs for the nestlings after they have hatched in an almost perfect match to the likelihood that all the eggs were fertilized by his sperm. And the cue he uses for this? It is the amount of time that the female was out of his sight during the time when she was laying the eggs and so could have been mating with other males. So elegantly simple a solution.

74. Does whom we marry matter to evolution?

In sexually reproducing species, each offspring inherits only half of each parent's genes (see Question 22). So why would you waste that perfect set of genes that you represent by splitting them up? The obvious solution is to choose a mate that is as similar to you as possible. However, if your mate is *too* closely related to you, you run the risk, under Mendelian inheritance, that some of your children will inherit two copies of any recessive gene for some deleterious condition that both of you happen to have. If you inherit just one copy of a deleterious gene, that is usually OK because its expression will usually be suppressed; but two copies is a disaster (see, for example, the case of sickle cell anemia) (see Question 68).

It turns out that genetically, cousins or second cousins are the best compromise: they are far enough away to minimize the probability of inheriting a deleterious recessive gene but close enough to share most of the rest of your genes. This is sometimes known as *optimal inbreeding* (or, alternatively, optimal outbreeding). In an elegant experimental study on quails, the evolutionary biologist Patrick Bateson demonstrated that individuals prefer cousins over more or less closely related birds of the opposite sex. There is even more convincing evidence from Iceland, where we can trace human pedigrees back through nearly 1,000 years: marriages between second cousins in this small, isolated population have been the most successful in terms of the number of descendants they have left—in effect, their genetic fitness.

However, the quality of one's genes isn't the only thing that is important for successful reproduction. If males defend feeding territories (or other resources) that the females can use to invest in rearing their offspring, then the quality of a male's territory may also be a consideration. This is especially common in birds, where territory quality can have a significant impact on offspring survival rates. In such cases, females may prefer to mate with the male with the best-quality territory and not just the best genes. If the range of variation in territory quality is large, females might prefer to be the second or even third female on a rich territory than the only female on a poor one. This is exactly what happens in the American red-winged blackbird, and the territory quality at which females switch their preference is known as the *polygyny threshold*.

We see this even in humans: polygamy tends to be more common where males have land or other forms of wealth that their wives can invest in rearing their children. In such societies, males with low status and little wealth are discriminated against by women when looking for husbands (as well as by families seeking husbands for their daughters); they rarely have more than one wife and, even then, they often have to settle for older women (i.e., those with lower reproductive potential) or women with disabilities or other social or physical disadvantages that might affect their reproductive potential, as the evolutionary anthropologist Monique Borgerhoff Mulder was able to show in her seminal study of East African Kipsigis agropastoralists.

The design of the mammalian reproductive system, with its much greater investment made by females because of internal gestation and lactation, means that female mammals should be much more choosy than males because they have more at stake on each conception. Analyses of people's stated preferences for partners in personal ads (or "Lonely Hearts" columns) confirm this: women are typically more demanding than men, and men will make more effort to advertise the qualities that women express an interest in. By contrast, unless they are rich, men tend to specify fewer demands, and women typically include fewer as descriptive qualities of themselves in their ads—at least until their declining fertility starts to have an impact on their attractiveness to men. It's an example of optimal foraging (see Question 12) at work in another domain.

75. Are humans really monogamous?

Most of us are familiar with those peculiar feelings and emotions that we describe as "falling in love"—a dreamy expression, a focus on one particular person to the exclusion of almost everyone and everything else (sometimes even to the extent of losing one's appetite), an overly positive view of the person concerned (the "rosy-colored sunglasses" phenomenon), and an overwhelming desire just to be near that person. Of course, not everyone experiences these feelings to the same extent, and some people may never experience them. But in general, falling in love is about as close as one gets to a human universal—and an example of how easily our biological emotions can overwhelm our celebrated rational capacities.

For humans, falling in love forms an important part of the process of pair-bonding, and is often (but not always) a prelude to a successful reproductive relationship. Indeed, studies suggest that the rosiness of the romantic sunglasses through which a partner is viewed at the outset of a relationship does predict how long the relationship lasts. Interestingly, psychologically speaking, it is not too dissimilar from the way people of both sexes react to charismatic leaders, including God and the saints.

Although it is certainly true that some romantic relationships can last a lifetime, humans are not strictly speaking monogamous. Set against the lifelong monogamy of species like the klipspringer antelope (perhaps the most monogamous of all the deer and antelope family) or the gibbons, night monkeys, and titis (undoubtedly the most monogamous of the apes and monkeys), humans make a poor showing. In some hunter-gatherers, where "marital" relationships can be quite informal, an individual can have as many as 10 or 12 partnerships during a lifetime, with children resulting from many if not most of these. This is probably not too different from what we observe in modern western societies where relationships have become much less formal in recent decades. In fact, the majority of societies allow polygamy—mostly for the benefit of men, but in a few cases (like the Tibetans, the Toda of south India and the Nymba of Nigeria) for the benefit of the women.

In this respect, humans seem to sit midway between the truly monogamous mammals and the truly promiscuous ones. This ambivalence is reflected in many indices that correlate with mating system. The degree of sexual dimorphism in body size, for example, is a good predictor of the mating system, mainly because males in promiscuous systems have to fight each other for access to fertile females, and that creates an arms race for large body size and ever bigger weapons. In strictly monogamous species, adult males and females are of approximately equal size, but in promiscuous species males are always much larger than females (gorilla males are twice the size of the females). Although human males tend to be larger than human females, the difference is modest—about 7% in stature and 20% in body weight. The size of the male testes is another good predictor of the mating system, with males from promiscuous species having much larger testes than males from monogamous species. Humans again lie midway between the two.

A more obscure index is the ratio of the second and fourth digits (the ring and index fingers) (the 2D:4D ratio). This ratio is about equal in monogamous species, but smaller than unity (longer ring finger) in promiscuous species (with both sexes exhibiting this pattern). Humans, again, lie somewhat in between these two limits (although this seems to be because both men and women exhibit two distinct types— one more promiscuous, the other more monogamous). This ratio is partly determined by the levels of testosterone that the fetus is exposed to in the womb—this being higher in polygamous species than in monogamous species—and, at least in humans, by the particular alleles of the vasopressin receptor gene (functionally, an antidiuretic hormone) that a male has. Both digit ratio and the vasopressin allele predict men's preferences for multiple sexual partners, and this is also true for women, at least as far as the digit ratio is concerned.

Even if humans are not lifelong monogamists, we can at least say they are pair-bonded with close, focused relationships that last for a few years. This raises an interesting issue, known as Deacon's dilemma after the neuroanatomist Terry Deacon who first pointed it out. Human romantic relationships exist within a large multimale/multifemale community, yet there is relatively little attempt by unpaired individuals to steal other people's *bonded* partners. In most other animal societies, this would rapidly descend into a free-for-all. That it does not, and pair bonds are respected most (albeit not all!) of the time, implies an unusual capacity to inhibit what are known as *prepotent responses* (in effect, the tendency that young children have to grab the biggest piece of cake at a party). The capacity to inhibit responses in this way so as to allow others a fair share is particularly dependent on the frontal lobes of the brain (the area that has evolved out of all proportion in humans).

76. Are conflicts of interests inevitable in evolution?

The selfish gene perspective tells us that every individual has its own interests at heart. This means that whenever two individuals meet up, a conflict of interest will not be very far away because each will always want to do what best suits its own interests. There are three well known evolutionarily important examples of this that all relate to investment in offspring. One is conflict between parent(s) and offspring over how much investment the offspring should get. Another is conflict between the sexes in how much to invest in their joint offspring. The third is conflict between the genomes over control of how the infant develops.

Once parents make a significant contribution to their offspring's development, it inevitably creates an asymmetry: offspring would prefer to be given more resources (whether this is milk or money) than their siblings so as to grow bigger and more competitive, whereas the parents would like to distribute their resources more equitably between their offspring (partly to maximize their own fitness and partly to hedge their bets against individual offspring turning out unsuccessfully). This is a simple consequence of Hamilton's rule (see Question 25): an offspring is more closely related to itself than to any of its siblings (unless it is an identical twin), so it should prefer its parents to invest in it rather than in any of its siblings. The parents, on the other hand, are equally related to all their offspring, so they will prefer a more equitable distribution. Infant (and maybe teenage) temper tantrums are a real-life consequence of this.

Despite this, parents may sometimes prefer to invest differentially in their offspring. Doing so allows them to maximize their fitness by ensuring the survival of their lineage and their genes. In Tibet, where land is poor and its availability very limited, all the brothers are obliged to marry one girl. They don't always like it (and nor does the girl), but custom demands it: in a context where there is no spare land, it avoids having to split the family farms into ever-smaller, uneconomic units in successive generations. Functionally, it is much the same as the European strategy of investing all the family wealth in one offspring, leaving the rest to fend for themselves. In some cases, parents may even deliberately manipulate the survival chances of offspring, either by infanticide (many Rajput families in India prior to the twentieth century) or by underinvestment. Peasant farmers in northern Europe in the eighteenth and nineteenth centuries practiced what historians refer to as the *heir-and-a-spare strategy*: it involves ensuring that you have only two surviving sons—one to inherit and one as a backup in case it dies. This was achieved by investing much less in later-born sons, who had mortality rates as high as 50% even before reaching their first birthday. This did not affect daughters, by the way: they were all valuable because they could be married off to other families (and were sought-after brides if they came from landed peasant families).

The second case is pair-bonded monogamy. In many ways, it is the ultimate example of cooperation: two unrelated individuals work together on a common task (rearing their offspring). Perhaps the most dramatic example of this is offered by the marmosets and tamarins. In these tiny South American monkeys, the male is wholly responsible for looking after the twin babies; the female has them only for 10 minutes at a time so they can feed, and even then the male may decide when they have had enough. In part, this unusual breeding system arises because the female traps the male into being monogamous: with the male's help, she can produce twin infants twice a year, whereas if the male opts to roam in search of new females to fertilize she would be able to produce only one offspring a year because of the burden created by lactatation and infant care. In effect, she makes him a fitness offer he cannot refuse. This is only possible, however, because these species are so small. In larger-bodied species where females produce only one offspring at a time, this strategy is not possible, and female are usually left to rear their offspring on their own.

In monogamous species both parties obviously have a vested interest in the outcome, but because mates are not related to each other (so kin selection doesn't apply) (see Question 25), the public goods dilemma (see Question 73) kicks in: there will always be a temptation to invest slightly less than your partner. Although monogamy is the norm in birds (because both sexes can contribute equally well to the business of rearing), males will even so sometimes try to run a second nest with another female if the environment is rich enough and if females are willing to shoulder the extra burden of rearing. By the same token, females always have an interest in copulations with other males if that gives them a better choice of sperm with which to fertilize their eggs. Extrapair copulations, or EPCs, account for 5–35% of fertilizations across bird species. Matings with "the male next door" account for as much as 12% of copulations in the otherwise obligately monogamous gibbon, and are estimated to account for 5–10% of all conceptions in humans.

The third example begins at the moment of conception. Because the two sexes differ in their reproductive interests, fathers will want to maximize investment in an offspring at the mother's expense (especially if there is a risk that he may have no more offspring with her), whereas the mother will want to ensure that the infant does not drain her so much as to threaten her own survival or her future reproductive opportunities. As a result, there will inevitably be conflict between their genomes as to which controls development (and especially the demandingness of the offspring). One likely explanation for genomic imprinting (see Question 22) is that it reflects parental conflict over control of the embryo and its development.

77. Are there sex differences in behavior?

There has been much debate as to whether there are sex differences in behavior or cognition. In fact, this has probably been the thorniest topic in the whole of psychology. In fact, in humans at least, the differences between the sexes in intellectual abilities are minimal. Males genuinely do seem to be better at spatial problems (and maybe, as a result, map reading), and may also be better at abstract tasks, while females are much better than males at handling social tasks, in language, and in dealing with offspring. But that's about it. The two sexes are equally good at most intellectual tasks.

Far more important are sex differences in reproductive strategies that are deeply based in the biology of mammalian reproduction. Because females bear the brunt of the reproductive investment in mammals and mammal babies are fragile, some mechanism is needed to ensure that their carer doesn't abandon them prematurely. Since this is much too important to be left to chance, it seems that the whole process is automated and beyond our control. The birth itself and subsequently breastfeeding trigger the release of oxytocin, and this is responsible for a deep and instantaneous bonding between mother and baby (see Question 13). Males do not experience anything like the same effect. In some species of voles, the males take no part in the business of rearing after they have fertilized a female, and often behave very aggressively towards other animals. Injecting oxytocin into these males' brains causes them to be more tolerant and willing to stay with the female while she is rearing her young.

Because the two sexes differ in their initial investment in expensive eggs versus cheap sperm, female birds and mammals usually make much more complex mate choice decisions than males. They have much more to lose if things don't work out, and they are invariably trying to find a balance between several criteria that often conflict (e.g., good genes versus the resources on offer for rearing). This is as true of humans as any other mammal (see Question 74).

In Darwin's original conception of sexual selection, males either display to females (and females choose among them: intersexual selection) or fight it out with each other (and females accept whoever is the winner: intrasexual selection) (see Question 6). Humans seem to pursue both strategies: males both display (through dancing, bragging, storytelling, joking, taking risks) and compete directly with each other (by fighting or scaring off the competition). At the same time, there is strong evidence for female choice in romantic relationships: women seem to make up their minds much earlier than men do, and then go for the chosen one relentlessly until he capitulates.

In wild horses, baboons, chimpanzees, and humans, a female's ability to rear her offspring successfully depends on the number of 'friends' (or allies) she has. Much of this points to the fact that women have much better developed social skills than men. Girls learn language, for example, earlier than boys do and are generally much more proficient at it in childhood. Women are also consistently better in their social skills and their ability to handle social situations. They are better than men at mentalizing skills (see Question 69) and as a result have larger social circles. They also have much more intimate friendships than men do. Perhaps as a result, they tend to work very hard at trying to maintain relationships even in contexts where this is physically difficult to do (e.g., when friends have moved away). In contrast, men seem to operate on an "out of sight, out of mind" principle: if their best friend has moved away, they simply find someone else to substitute. Women are, however, rather less forgiving of transgressions than men. Since social skills are rarely considered in the (often rather acrimonious) debates over sex differences, these rather striking social differences are almost always overlooked.

78. Why do mating systems differ between species?

Many species simply dump their fertilized eggs in the environment and leave them to get on with developing into adults on their own. Other species invest more heavily in rearing so as to maximize the chances that each offspring will make it through to adulthood (see Question 59). This is usually, but by no means always, done by females. In some fish, for example, the male tends the developing eggs in a nest (sticklebacks are an example) or carries the eggs, and later the developing young, in his mouth (cichlids do this), or even grow them internally and give birth to live young (as the seahorses and pipefish do). The male midwife toad carries the fertilized eggs embedded in the skin of his back until they hatch as tiny toads. Even some birds have male-only nursery care: in ostriches, some stints, sandpipers and jacanas, females lay their eggs in the nests of several males, and the males do all the caring. Famously, the emperor penguin male huddles its one giant egg on his feet covered by a special belly flap through the Antarctic winter while the female goes off to enjoy the warmer waters off southern Africa to return several months later to share the childcare after the egg has hatched.

Although some bird species (such as chickens and ducks) have young that can feed themselves as soon as they have hatched (they still need to be guarded against predators, however!), many birds and all mammals have to feed their young to begin with. For most bird species, this involves ferrying insects or fish back to the nest, and both sexes usually share this duty. However, a few birds (the pigeon family) and all the mammals feed their young on milk produced internally. In pigeons, both sexes produce "crop milk" by sloughing off the lining of the throat and regurgitating it to their young. In mammals, of course, only the female produces milk through the highly specialized mechanism of the mammary glands. Because this can only be done by one sex, most female mammals rear their young unaided. However, in a few species, males may help the female by carrying the young (marmosets and tamarins) or defending a feeding territory. When males do help the female in some way, the result is almost always monogamy. Otherwise, the mating system defaults to some form of promiscuity or polygamy in which males defend clusters of females.

Because of these differences in the two sexes' abilities to rear the young, 95% of mammal species are polygamous, whereas around 85% of bird species are monogamous because both sexes can feed the young. The dog family (wolves, foxes, coyotes, etc.) is the one exception in mammals: every single species is monogamous, but that's because the male provisions the female and the pups with half digested meat that he brings back from hunting and regurgitates as half-digested weaning food at the den. That's something that works for meat, but not for vegetable foods.

The direction in which animals are pushed in this respect depends on how much parental investment the offspring need (mostly a reflection of how big their brains are: see Question 85), how easily the male can contribute to the business of rearing, and how the females distribute themselves in space. If predation risk is low, females may prefer to space out on their own (see Question 82), and males may be forced into being monogamous if providing a service for the female (e.g., defending mating access, protecting offspring from infanticide) makes it worth their while sticking with the female. If there is no such advantage, the male is better off defending a territory that contains several females and mating polygamously. If predation risk is high, females are likely to clump together for protection (see Question 81) and males are then likely to associate with the female group and defend it against rival males. If the number of females gets too large, the male will be unable to prevent other males joining the group; in which case, he will switch from defending the group to defending individual females as they come into estrus, leaving any other females that happen to be in estrus at the same time to mate with the other males. Once again, individuals are trading off between opportunities and constraints on what they can do.

79. Is human behavior always adaptive?

There has been a long-running debate as to whether human behavior in the modern world is adaptive, in the sense that it is designed to allow us to maximize our contribution to the species' future gene pool. Those who claim that it is not adaptive point out that, in the modern world, people voluntarily curb the number of offspring they produce or sometimes even decide to not have any children at all. This looks suspiciously like a form of genetic altruism. An alternative claim has been that the evolution of culture has allowed us to change our environment so radically that our genes have not had time to adapt (the "stone age mind in a space age body").

There are several things wrong with these kinds of claims. First, just because people produce fewer offspring than they could do does not imply that their behavior is not underpinned by Darwinian natural selection. The claim fails to appreciate that all evolutionary processes involve complicated trade-offs, as both Lack's principle and Hamilton's concept of inclusive fitness (Question 25) remind us. Evolution is not just a matter of pumping babies out as fast as you can; it is about maximizing the number that make it through to adulthood and breed in their turn. Fitness is about the number of great-grandchildren that you have, not the number of children. The fact that individuals face different contexts (especially in socially complex species) means that there may be several alternative, equally good ways of solving the problem of how to maximize one's fitness. A much more sophisticated analysis is required than those produced by these kinds of critiques.

The Cheyenne, one of the North American Plains Indian tribes, provide an example. They had two kinds of chiefs: hereditary peace chiefs, who ran the society, and war chiefs, who defended the tribe against raiders. War chiefs took an oath of celibacy and vowed never to leave the battlefield alive unless they won. However, if they survived, war chiefs could be released from their vows and could marry. When they did so, they were often an attractive catch, probably for precisely the reason that Zahavi identified in his handicap principle (see Question 6). On average, peace chiefs produced 3.8 offspring, whereas war chiefs had only 2.6 offspring. However, the variation in reproductive output was much greater for war chiefs because many of them died in war and never had the chance to marry. In fact, those who survived the battlefield were reproductively much more successful than most peace chiefs. They took a bigger risk for a much bigger prize.

However, there is something else we need to know about warchiefs. Whereas peace chiefs inherited their title, most war chiefs came the bottom of the social scale and suffered considerable discrimination. Their choice was between taking the risk of being killed or being socially oppressed and being very unlikely to marry because they had nothing to offer their brides. The only way orphans could hope to compete with their better-off rivals was to take enormous risks. War and peace chiefs represent a classic high-risk/high-payoff versus low-risk/low payoff strategy: the peace chief route was undoubtedly the safest, but it was not an option for orphan males. What makes this example particularly instructive is that the demographic records show that the payoffs and costs of the strategies mirrored each other: the two strategies were in evolutionary equilibrium (or what biologists refer to as an *evolutionarily stable strategy,* or ESS). You could do equally well either way, but which option an individual preferred depended completely on their circumstances.

Another frequent claim is that culture has changed our environment so that our genes and the behavior they drive are out of kilter with our circumstances. This one is actually much simpler to deal with. The behavior of most large mammals (and humans in particular) simply isn't that tightly determined by their genes. The point about having a big brain is to be able to adjust one's behavior on a day-by-day basis according to one's changing circumstances. In reality, most of the contexts in which we make our reproductive decisions haven't actually changed at all since the Pleistocene era. We still compete for mating partners who differ in quality; we still make decisions about how best to invest in our children and grandchildren; and we still make decisions about which foods to eat and which not. The circumstances in which we do this may have changed, but not the consequences of our choices or our ability to make efficient decisions.

The point is that it is not our behaviors that are genetically inherited; it is the capacity (or brain power) to make decisions and the motivations (or goal states) that provide their evolutionary objectives that are genetically determined (see Question 80). How we choose to achieve those goals, and which behaviors we use to do so, depend on how we assess the costs and benefits involved. Many of the skills we need for doing this have to be learned and that is why primates have a prolonged youth.

It is also important to remember that natural selection never results in perfection as such (see Question 18) but rather in the best that can be done under the circumstances. Nor does it mean that everyone will be reproductively successful. Some individuals will inevitably make bad decisions, but that is part and parcel of the business of natural selection: without that variation between individuals, there would be no natural selection, and no evolution.

80. Doesn't the theory of evolution imply that there is no such thing as free will?

Ah, the question that always comes up eventually and the one that reveals the bugbear that underlies most concerns about evolution. The short answer to this question is simply: no. As we saw in the previous question, the point of paying the enormous cost of having a big brain is to be able to make your own decisions about what to do in light of the particular circumstances you happen to encounter. It is worth having everything determined by your genes only if your circumstances and the decisions you need to make never change. That is probably ever true only for bacteria and viruses. For the rest, it helps if the organism has some control over its own destiny so that it can fine-tune its behavior to the circumstances of the moment.

Natural selection simply sets the ground rules that determine the costs and benefits of alternative courses of action. It usually does that by establishing some goal states in the organism ("always keep your energy levels up") with some kind of signal to let the body monitor its state (hunger pangs when energy levels are low) and a motivational system that allows you to fulfill that goal (find something to eat when the hunger pangs kick in). It is important that the organism has control over how to fulfill the goal, because its circumstances won't always be the same. It is then up to the individual to do the best it can. If you choose well, you will leave many descendants; if you choose badly, you won't.

Of course, we all have innate predispositions, or prepotent actions, that, all else equal, make us prefer one course of action rather than another. But for large-brained species, learning becomes an increasingly important part of that mix as a way of moderating those prepotent actions—allowing us to see both the short- and long-term consequences of our actions, and providing us with the basis for deciding which strategy is the best among our available options (see Questions 60 and 78). If we grab the whole cake, that may satisfy our short-term craving for sweet food, but it may cause us to pay a longer-term cost because we upset everyone else when they get none—or we overdose on sugar and make ourselves ill. If we destabilize our friendships with everyone else, we risk leaving ourselves without an ally on some future occasion just when we need their support. We have to learn to recognize those consequences and be able to adapt our behavior accordingly. Without that ability, we will be dead in the evolutionary water. And extinct.

There is an important reason for being able to do this: the kinds of societies that we and other primates (monkeys and apes) live in are implicit social contracts—we club together to gain some greater benefit (see Question 81)—but for that to work we have to be able to inhibit our natural tendency to satisfy all our desires irrespective of the consequences for everyone else. If we can't, our social systems will break down (as they demonstrably do when things get out of hand during civil wars and there is a loss of civic control). This is mainly why we have to teach children how to behave morally and courteously toward others. They do not do it naturally, despite some well-meaning attempts to claim that they do.

Of course, this is not to say that everyone is a perfect citizen. Some people never learn to behave in this way, either because of the circumstances in which they grew up or because they lack the ability to inhibit their own behavior—the two often go together, the one reinforcing the other. Long-term studies of individuals from childhood to adulthood suggest that individuals who fail to learn these skills as children continue to behave antisocially as adults, with boys being more prone to this than girls.

Ultimately, however, it is we who choose to behave the way we do. We are not driven to do so by our genes, even though these may predispose us to behave in certain ways. As most religions have recognized, we always have the final say in deciding whether or not to give way to the "temptations of the flesh."

9

EVOLUTION OF SOCIALITY

81. Why do some animals live in groups?

The ancestral mammals were small (we can tell that from their fossils) and almost certainly solitary (as suggested by reconstructing their likely behavior from the social arrangements of living species using statistical analyses that take species evolutionary history into account). Living in social groups arose later in some, but not all, lineages and could have evolved for a number of different reasons.

One reason for forming groups is that animals simply gather together in temporary aggregations on good feeding sites. Many flocking birds, such as geese and shorebirds, are examples. These groups are rarely stable, being simply groups of convenience (*aggregations*). Animals join and leave as a function of how rich the foraging is at any given time. Such groupings contrast with *congregations*, where animals remain in a stable group for some considerable time, if not for life, moving together as a coherent group from one feeding site to another. These groups typically exist because the members gain an advantage from grouping permanently together.

One obvious reason for forming stable groups is to facilitate reproductive cooperation, usually in the form of reproductive pairs. Such an arrangement can be temporary and may be confined to the breeding season (*facultative monogamy*), with the animals living solitarily or in loose flocks the rest of the year (some voles, many of our familiar garden birds) or they may be permanent (parrots and eagles, as well as some primates: life-long or *obligate monogamy*). In some cases, monogamous pairs may gather together in large, relatively anonymous herds or flocks. Examples include most colonially nesting seabirds or species like sand martins and bee-eaters that nest in sandy riverbanks or cliffs. In these cases, the large aggregations may be due to the limited availability of suitable refuges, combined with the protection from predators provided by the group.

There are two main non-reproductive reasons for living in stable groups. One is cooperative hunting, whereby several individuals cooperate to capture much larger prey than any of them would be able to do on their own. Lions and hyenas are two obvious examples. In most cases, these groups are quite small (perhaps 5–15 adults). Chimpanzees cooperate (to some extent at least) when hunting monkeys, but in this case there are good reasons to think that this is a derivative benefit rather than the main factor selecting for group living. Aside from the fact that hunting is actually very rare in most chimpanzee populations, their groups are very large (50 animals or more) compared to the number of individuals actually involved in a hunt (usually half a dozen males at most).

In fact, defense against predators is probably by far the most common reason for animals to live in groups. In primates, group size correlates with the predator-riskiness of the environment, and this has also been shown to be the case for some antelope species that live in permanent groups. One indication that predation is a serious problem for primates is indicated by the fact that groups will bunch up and travel faster in areas where there have been recent attacks by predators or when predators are encountered. In most cases, groups simply deter predators rather than actively driving them away (see Question 83). Some species (beavers, corals, and social termites) occupy communally constructed defensive structures within which they can live securely.

All these benefits arise through *group-level* (or *group aug-mentation*) *selection*, where individuals have higher fitness by cooperating together than they would by living alone (see Question 73). An obvious question, however, is why, if social groups have evolved to fulfill an essential function for animals, they do not live in the largest possible groups. Most such species have groups that average just 10–20 animals, and none average more than about 50 animals. In contrast, herds of up to a million animals were said to have occurred among American bison in the nineteenth century, saiga antelope in Russia, and wildebeest in East Africa—all species that live in unstable, anonymous herds within which individuals do not have regular relationships (aside from mothers and their dependent offspring, of course).

82. Why can't groups become infinitely large?

The answer is: because of a classic evolutionary trade-off. A selection pressure (usually the need to minimize predation risk) favors ever larger groups, but the costs of living in large groups eventually limit their size. These costs have four distinct components: time, food, fertility, and cognition.

Although it has often been overlooked by ecologists (who have tended to focus exclusively on habitat richness and energy flow), time is a major constraint for most large animals because they have to be able to do everything they need to do (feed, move, rest, socialize) within a defined physiological time cycle (usually 24 hours). These time demands reflect the foraging quality of the environment, but they also reflect other environmental demands placed on the animal's physiology. An example of physiological constraints would be the time needed by ruminant species (cows, antelope, deer) to ferment and digest the food that has been eaten.

If the vegetation is of poor quality such that nutrient extraction rates are low, animals will need to spend more time feeding in order to meet their daily nutritional needs. As a result, they will have to spend more time traveling because they quickly exhaust food patches and need to move on to find a new one. In addition, environmental parameters, such as air temperature, may force animals to rest when temperatures are at their highest (usually around midday).

Species that live in bonded social groups face an additional constraint: the time that has to be invested in social grooming to create these bonds (see Question 86) is considerable and increases with group size. We humans devote about a fifth of our waking day to social interactions for this purpose. If relationship quality depends on the time invested in it (as is the case in monkeys, apes, and humans), then the time available for social interaction (after accounting for foraging and any downtime imposed by the environment or physical activities) will limit how large a stable group can be.

Breaking through this glass ceiling requires a way of using time more efficiently. This could involve a change in diet or digestive physiology, an increase in body size (to exploit efficiencies of scale), a reduction in travel costs, or a more efficient method of social bonding. Each of these is likely to be associated with its own costs, creating a feedback loop. An increase in body size, for example, will require more energy to fuel the larger body, and that means more time devoted to feeding and traveling (unless you change to a more energy-rich diet). Nothing in biology comes without a cost.

The third constraint is the effect that living in groups has on female fertility, in particular. Ecologists have tended to assume that the costs of living in groups are always ecological (competition for food and the safest resting places), but in reality these are trivial compared to the fertility costs of group living. The endocrinological system that underpins the menstrual cycle in mammals is very finely balanced and can easily be disrupted by both physical and psychological stress. This is a by-product of the mechanism that switches off the menstrual system during pregnancy and lactation (the phenomenon known as postpartum amenorrhea). It seems that this system can easily be suppressed by stresses incurred from living in groups as individuals bump into or bully each other.

Of course, you could tweak the system's sensitivity to solve this, but that would affect the mechanism (amenorrhea) that ensures that the mother has to feed only one offspring at a time. For large-brained species with long interbirth intervals like monkeys and apes, any delay incurred in initiating the next reproductive cycle means a significant reduction in fitness. Most large-bodied primates, for example, produce only around five offspring in a lifetime, so losing even one offspring represents a very significant proportion of their fitness.

This stress effect results in a negative relationship between the number of females in the group and their fertility across all mammals. When all else is equal, this effect appears to place an upper limit on the size of groups at around 5 or 6 breeding females (and hence a total group size of around 15–20). The obvious alternative is to live in flexible fission-fusion social systems (as herd-forming species do), since this allows individuals to leave when the stresses become too great and rejoin later when they are lower. That option, however, is not available for species that live in bonded societies, where leaving and joining are usually resisted by group members actively repelling strangers that attempt to join. Those few species like monkeys and apes that live in larger groups have solved this problem by forming coalitions that buffer the females against these stresses. However, this does not remove this cost completely: it simply defers it to larger group sizes.

Finally, at least for intensely social species like monkeys and apes, group size may also be constrained by the animals' ability to manage their social relationships and, to the extent that this is dependent on cognitive abilities, by the size of their brain (see Question 85). They can get around this problem only by growing a larger brain. But brains are energetically very expensive (see Question 71), and that means eating more food: diet and food availability determine whether the species *can* afford to grow a bigger brain, and that brings us back to time. It is these feedback loops that create most of the costs that make evolutionary change an uphill struggle for all species: benefits do not come for free in the biological world.

83. How do large groups avoid the public goods dilemma?

In answering Question 82, I spoke as if stable social groups are a cooperative venture. If this really were so, they would fall afoul of what economists refer to as the public goods dilemma (see Question 74). They would be at risk of freeriders who took the benefits of the cooperative venture but didn't pay all the costs. Since this is a form of genetic altruism, such behavior should be selected against. If the frequency of freeriders becomes too high, the group will disintegrate because the conscientious individuals will become reluctant to shoulder the burden that allows the freeriders to gain a benefit from doing nothing (see Question 73). This would set a very low limit on the size of social groups and might limit them to monogamous pairs (whose only function was reproduction) or small groups of cooperative hunters—in other words, groups with a very limited, immediate purpose.

If protection against predators were based on active defense by group members, then the freerider problem would certainly arise: those who hang back and allow others to take the risks of squaring up to the lion would benefit from the altruism of the active defenders. In fact, most anti-predator strategies are passive: by simply grouping together, animals deter a predator from bothering to attack. This is because predators prefer to attack isolated individuals, partly so they aren't distracted by other animals running in front of them and partly because most predators need an element of surprise if they are to be able to capture their prey. Each predator species has its own preferred attack distance, reflecting its style of hunting, how good it is at using cover, and its speed from a standing start. Groups that have more pairs of eyes are more likely to spot them coming and so have time to take evasive action. Many deer and antelope species, for example, have a white underside to their tails that they flash at a predator as they run away to let it know that it has been spotted (a kind of "there's no point in launching an attack now" signal).

A great deal of ink has been spilled trying to find a solution to the evolution of behavioral cooperation that doesn't fall afoul of the public goods dilemma. However, it seems that all this effort may have been wasted, because most social groups are not cooperative ventures in this sense. For species like primates that choose to live in stable social groups for protection, the problem they face is actually a coordination problem rather than a cooperation problem. Cooperation is always subject to the public goods dilemma (the temptation to work less hard for the common good) (see Question 74), but a coordination problem is not. You are either in the group or you are not, and the only way you get the benefit of the group is by being in it. There is no up-front entry fee to pay (which is what creates the public goods dilemma). Instead, all you have to do is be willing to stick with everyone else (i.e., coordinate your movements with them).

Coordination of this kind certainly incurs a cost, but it isn't a cost you can make others pay on your behalf. The cost is a disrupted time budget: you may have to travel farther because a larger group needs a bigger range to allow everyone to find the food they need, and you have to be willing to stay with the group when it rests or moves even if you would prefer to do something else. That doesn't come cheaply. It requires intense bonding that motivates you to stay with your friends or family and the cognitive ability to appreciate the consequences of your actions, as well as the capacity to inhibit actions (such as wandering off when everyone else wants to rest) that would destabilize the group (see Question 78). In effect, you need to be able to trade short-term losses for long-term gains, and this cognitively difficult calculation may explain why only species with large brains have these kinds of groups.

This isn't a problem for those species that form fission-fusion societies, because individuals can join and leave as it suits their momentary interests. This is the main reason why herd-forming deer and antelope exhibit sexual segregation outside the mating season. Because males are typically much larger than females (their mating systems tend to be contest-based matches in which males fight each other), they take much longer than females do to fill their much larger stomachs before needing to ruminate; as a result, males carry on feeding long after the females have gone to rest and so gradually drift off, leaving the females behind. The small antelope that have pair-bonded social systems are monomorphic (males and females are the same size), and that reduces the need for one sex to carry on feeding when the other wants to rest.

84. What's so complicated about social life?

While it is true that many of the same rules about the costs and benefits of joining and leaving groups will be common to both casual herds and bonded groups, there are some important differences between aggregations (casual groups that form for some immediate benefit and disperse again as soon as the benefit ceases) and congregations (stable social groups that maintain a level of social cohesion over time even when there is no immediate benefit).

Aggregations (or fission-fusion social systems) have the advantage of minimal cognitive cost, and the benefit of considerable flexibility that allows the ecological and physiological costs of group-living to be dissipated. But they suffer from a lack of certainty: the risk is that you might find yourself with no one nearby when the one predator for miles around happens to turn up. Worse still, even if you do happen to have companions, there is no guarantee that they will stay with you when the predator appears. Most of these herding species follow an everyone-for-themselves strategy rather than an all-hands-to-the-pump one. Living in a stable group incurs significant costs (being tolerant of others, having a sense of commitment to others, coping with disrupted time budgets) that require the evolution of a large, very costly brain to manage. But their advantage is certainty that others will always be nearby because they won't let you drift away without noticing, as is so often the case with casual aggregations. The issue is that these kinds of relationships have to be set up way ahead of the time when you need them.

If predation risk increases and a larger group is needed, then some solution also has to be found to mitigate the fertility costs that living in groups incurs (see Question 82). Group size can then increase, but only up to the next glass ceiling, because there will always be one. Indeed, the distribution of group sizes in primates turns out to consist of a series of very specific values that seem to correspond to a set of natural optima or glass ceilings. The way primates, and perhaps a small number of other mammals (including the horse family and the elephants), have solved the fertility problem is by females forming protective coalitions, either with each other or, in a small number of cases, with males who in effect act as "hired guns" or bodyguards (examples of the latter include the gorilla and the gelada baboons).

These coalitions, built around grooming bonds, are essentially a balancing act: they work mainly by keeping everyone else at a distance so that they don't impose stresses by coming too close, but at the same time without driving them away so completely that they leave the group (which would mean losing the benefits provided by the group altogether). These kinds of relationships are complex to manage. In effect, they require the capacity both to forecast the future consequences of one's actions (and evaluate any alternatives) and to inhibit actions that would produce a short-term gain at the expense of a much larger long-term cost.

In larger social groups, the formation of coalitions creates a layered structure to the social network of the group: instead of everyone interacting with everyone else, each individual confines its interactions to its core coalition partners. Its remaining relationships with other group members are virtual and can be dealt with third-hand (e.g., by observation of third-party interactions, by reputation management, or by inferences about who is related to whom). Maintaining virtual relationships requires us to model their intentions in a virtual world because this is something we know about only indirectly by inference from observed behavior. In humans, this form of modeling the relationships behind the behavior in the mind has been shown in neuroimaging studies to be cognitively much more demanding than manipulating factual information about the physical world: it requires the recruitment of many more neurons.

85. Why do some animals have very large brains?

There has been steady pressure to evolve larger brains in certain families of animals. Monkeys and apes, elephants, the horse and camel families, and the toothed whales (dolphins and porpoises) have all increased brain size (both absolutely and relative to their body size) over geological time. Others, such as the cat family, the dog family, and the deer and antelope, have shown almost no change in relative brain size across their entire evolutionary history. The families that have evolved larger brains are all characterized by living species that have bonded social relationships. They have groups in which individuals have, for want of a better word, close friendships with each other. There are few, if any, similarities in any aspects of their ecology.

This has some parallels in the small-brained species in that the dog family (which includes wolves, coyotes, jackals, and foxes as well as our more familiar family pet) has bigger brains than the cat family. The reason is that all the dogs are monogamous (i.e., have lifelong pair bonds), whereas all the cats (with the sole exception of the bigger-brained lions) are solitary. Pair-bonded relationships are complex phenomena that require a lot of orchestration and therefore a bigger computer to manage effectively. We see the same effect among the antelope, the bats, and the birds: monogamous species always have larger brains than promiscuous species that do not form substantive relationships. The monogamous birds illustrate this particularly well: species that pair for life (parrots, eagles, the crow family) have larger brains than species that find a new mate each year (most of the smaller garden birds) because maintaining a successful lifelong relationship is cognitively much more taxing (*as we all know*), and both in turn have larger brains than species that mate promiscuously.

Only a small number of families (monkeys and apes, the horse family, elephants, dolphins) have groups larger than pairs that are based on bonded relationships. Their groups typically vary between 5 and 100 individuals in size, but are never as large as the casual herds found in many large-bodied ungulates. In primates, the monogamous species have the smallest brains because they have only one bonded relationship to look out for, whereas species that live in bigger groups have much larger brains because they have to manage proportionately more relationships. Among the monkeys and apes and the whales, social group size is strongly correlated with brain size, a relationship known as the *social brain hypothesis*.

An alternative hypothesis might be that animals need large brains to enable them to forage in more sophisticated ways. Foraging skills do indeed correlate with brain size, as has been repeatedly shown in primates, but since social group size does so too, this should prompt us to ask whether both are independently selecting for large brains or whether they have some other configuration (e.g., one is the cause and the other a constraint or a consequence). And herein lies an object lesson about how we test evolutionary hypotheses. If the latter is in fact the case, but we treat it as the former, then we risk confusing Dobzhansky's two ways of testing for adaptations (outcomes versus process) (see Question 12). When we undertake such tests, we need to be absolutely sure that we are comparing like with like; otherwise, we get what is known as GIGO (garbage in, garbage out) science.

In fact, if we do the analyses correctly, it turns out that, in living species, foraging skills constrain brain size, and brain size constrains group size. From an evolutionary perspective, of course, the causality is reversed: the need to live in large groups has selected for large brains; and if you want a large brain, then you need to solve the problem of how to acquire the extra nutrients needed to grow and maintain a large brain. That may well mean having to change your diet to one that provides more nutrients per mouthful (such as switching from a leaf-based to a fruit-based diet) or foraging more smartly (e.g., being able to circumvent plants' defenses [see Question 52] or learning how to predict fruiting cycles).

It is important to be clear that the social brain hypothesis *is* an ecological hypothesis. It identifies an ecological problem as the main driver, with sociality as the solution. Living in groups is not an end in itself—it incurs far too many costs (see Questions 83 and 84) for that ever to be the case. The crucial question here is whether animals' survival and fitness is most strongly affected by finding food or by predation, and then, whichever is the case, whether they solve this problem by individual trial-and-error learning or socially through other group members. The social brain hypothesis explicitly identifies predation risk as the principal driver, and living in groups as the solution to that problem (see Question 81). It then identifies the maintenance of group integrity through time and space as the principal problem that animals have to solve in order to be able to achieve that evolutionary objective, and a large brain as the solution to *that* problem. And then because large brains are expensive, it creates a foraging problem that needs to be solved. The demands of foraging thus act as a constraint on the capacity to evolve a large brain, not the selection factor favoring a large brain.

The social brain hypothesis makes a very specific prediction about the natural size of human social groups based on our brain size (that they should be about 150 individuals). This prediction has been widely confirmed by ethnographic, historical, and sociological evidence on the size of both personal social networks in modern societies and community sizes in traditional small-scale historical and contemporary societies. In addition, around a dozen neuroimaging studies have shown that in both humans and monkeys, the social brain hypothesis holds *within* species as well as between species: individual differences in social network size correlate with the volume of key brain regions, with the regions involved in mentalizing (see Questions 69 and 84) being especially important.

Large brains are needed to manage the number of relationships in large social groups, partly because the number of third-party relationships ("friends of friends") escalates very quickly with group size and partly because of the need to maintain coordination and prevent the group from drifting apart (see Question 82). These are cognitively difficult tasks that require skills of diplomacy that are not needed in simple fission-fusion herding social systems. Indeed, they are so difficult that a prolonged period of social learning prior to adulthood is needed in order to learn and practice them. This seems to be the main reason why primates have such long periods of social development (in humans, taking up most of the first 25 years of life).

86. How do animals bond their groups?

In primates (and perhaps species like horses and even elephants), the bonded relationships that make stable social groups possible are usually serviced through some kind of social grooming or other forms of physical contact. We know from experimental manipulations (in monkeys) and brain imaging studies (in humans) that endorphin activation lies at the heart of social grooming: the brain's endorphin receptors go crazy when we are being stroked. This effect is mediated by a highly specialized neural system, the afferent C-tactile (or CT) system. This system is very unusual: its receptors are found mainly in the hairy skin, and its afferent neurons are unmyelinated (so transmit very slowly), respond only to light, slow stroking at a rate of exactly two centimeters (about one inch) per second (about the natural speed of hand movements during grooming), and, unlike all other peripheral nerves, have no return motor loop (the ones that normally cause you to pull your hand away from the fire when you feel the pain).

Endorphins create an opiate-like sense of relaxation and a feeling of contentedness (and hence trust) with those with whom you engage in this activity. The endorphin connection probably evolved originally to facilitate mother-offspring bonding in mammals. Most mammals actively lick their off-spring as well as cuddle them, and this calms infants. This is also why human infants calm when rocked: it turns out that the inner ear is full of CT fiber receptors that are activated when the head is moved in a regular way. It seems that this effect was co-opted in highly social species to manage bonded relationships among adults when these evolved later, perhaps because a sense of trust emerges naturally out of a sense of relaxation and calmness induced by the endorphin surge trig-gered by social grooming.

Endorphins are in fact a key component of the brain's pain management system. So it is perhaps no surprise that they are also activated by physical exercise. Indeed, they are the pri-mary explanation for "runner's high" (or "second wind")—the point in a long race when the stresses on the body suddenly seem to drop away and feel you can coast along effortlessly forever. Because social play involves both intense physical ex-ercise and social contact, endorphins may also be responsible for the pleasure of social play.

Because grooming is so intimately involved in social bonding in primates, the amount of time devoted to grooming (back to time again) increases in line with social group size. However, this is not because you have to groom with every other individual in the group. Rather, it reflects the fact that individuals are investing more heavily in their core social partners, their allies. The number of core grooming partners does not increase with group size, but the pressure and stress created by other individuals does, and this requires a proportionately more effective coalition to counteract these stresses. The fact that animals groom more with their close allies in large groups seems to reflect the need to ensure that the bonds with these individuals are as strong as possible so that they will function effectively on that one occasion when they are really needed. Other relationships in the group are dealt with virtually, via a kind of friend-of-a-friend process, and through knowing others by sight and by reputation (see Question 84).

There is considerable evidence from primates to show that a female's fertility, her own longevity, and the likelihood of her offspring surviving to adulthood all correlate with the number of friends she has. Even wound healing is faster for those who have more friends. In other words, for these intensely social species, a crucial factor determining a female's fitness is her social network, and hence the social skills she uses to build and maintain this. This is not the case for males, whose fitness is typically more dependent on their ability to compete for mates.

Much the same is true of human social relationships. The amount of time devoted to a friend correlates with how emotionally close we feel to them and to our willingness to help them out. We devote 40% of our total social effort (whether measured as time or emotional capital) to just five individuals, the ones who function as our support clique (the individuals on whom we depend for emotional, social, and financial support). And, as with other primates, the number of close friends we have significantly affects our health, well-being, and even longevity.

The time animals have available for this kind of social interaction is, however, limited by the demands of their other requirements (see Question 83) and this, as much as anything, sets the limit on social group size in monkeys and apes. Species like the folivorous colobus monkeys have to devote large parts of their day to resting because fermenting the leaves they have eaten requires complete rest (as it does when cows ruminate). Since this is incompatible with anything else, the time they have for social interaction is very limited, and they are constrained to live in small groups of 10–15 individuals. Frugivorous monkeys like the baboons avoid this problem because nutrients can be more easily extracted from fruits; as a result, they have more time for social interaction (and play) and so live in much larger groups. Even so, the time they have for social interaction limits the typical size of their groups to around 50 individuals.

87. So how do humans bond their much larger groups?

If the time available for grooming limits the groups of the most social monkeys and apes to around 50 individuals, then how on earth do we humans manage to bond our groups of 150 friends and family—never mind the mega-communities that we now live in?

We know that humans use the same bonding mechanism, because brain scan studies have shown that slow stroking, such as occurs in grooming, triggers an endorphin surge in the brain. We also know that people's attachment style (how warm or cold they are in their relationships) correlates with the density of endorphin receptors, especially in the brain's frontal lobe. People who are cool and distant in their relationships have fewer receptors. It is as though they fill these up quickly and then don't seem to want to have any more relationships. We also know that the number of friends people have correlates with their absolute pain threshold (itself an index of endorphin receptor density).

Because grooming (like cuddling in humans) is a strictly one-on-one activity, it is physically impossible to groom with two individuals simultaneously with equal intensity. (If you don't believe me, try it. I can guarantee one of them will get offended because you aren't paying them enough attention.) This sets an upper limit on the size of social group that can be bonded through social grooming, and this limit is around 50 individuals. To break through this glass ceiling in order to be able to bond groups of 150, our ancestors had to find ways of triggering the endorphin system that did not require physical contact. If this could be done at a distance, then several people could be "groomed" simultaneously. Over the course of the last two million years since our social group size first breached the 50 limit, it seems that we have found a number of ways of doing this.

These have included laughter, dance, singing (without words), emotional storytelling, and many of the rituals of religion, as well as communal eating and the consumption of alcohol. These all turn out to trigger the endorphin system (sometimes better than grooming does, even). Each of these has the advantage that we can trigger endorphin activation in others (as well as ourselves) without actually having to touch them. That means that the number of people we can groom simultaneously can increase dramatically, allowing us to use what social time we have much more efficiently. Laughter groups (the number of people who laugh together on a social occasion) are similar in size to conversation groups (around four people); dance groups are around eight; singing groups can be much larger, perhaps several hundred people.

Of these, laughter seems to be the oldest (it is the most visceral and uncontrollable, and we share it with the great apes). It probably first appeared some two million years ago when the genus *Homo* first emerged. Singing and dancing were probably added to the toolkit piecemeal much later after the appearance of archaic humans half a million years ago as larger groups were needed. Feasting and storytelling, along with fully modern language, may have come later with the appearance of modern humans.

88. How can we explain the evolution of deception?

When the social brain hypothesis (see Question 85) was first proposed, it was in fact conceived in terms of animals' abilities to outwit each other so as to steal others' food and resources. It was this sense of political scheming that led to its being named the *Machiavellian intelligence hypothesis*, after the infamous late medieval Italian political philosopher Niccolò Machiavelli.

The problem with this view is that theft and deception are destructive of social life and inevitably lead to the dissolution of groups rather than their stability. It is a form of freeriding: I do all the hard work of digging up some nice juicy root, and you steal it off me. In other words, it would be yet another cost to group living that would limit group size in the face of whatever benefits the group offered, much as the central problem of fertility does (see Question 82).

The real issue is that you can exploit someone only if you actually live with them, so group living must have evolved before these kinds of Machiavellian behaviors. In neither evolution nor everyday psychology can you invite someone to join your group so that you can exploit them; but you may be able to exploit them once they are in your group. And if Machiavellian behavior does require bigger-than-average brains, then being in a group that depends on a large brain for its stability makes it a natural evolutionary development.

There is, however, an important distinction between antisocial (or selfish) lying and prosocial lying. The difference lies in the intentions of the liar: antisocial lying aims at gaining an advantage at someone else's expense, whereas prosocial lying is concerned with repairing relationships in contexts where a relationship has been, or might be, weakened. Examples of the latter include saying how nice someone looks when in fact you are unimpressed by their new haircut or dress (but you don't want to upset them by saying so), covering up minor indiscretions or omissions (to avoid causing an unnecessary argument), and, perhaps, "liking" an online post because you don't want to upset your friends. In other words, it is all about neutralizing the kinds of minor, everyday things that could easily be misunderstood and inadvertently damage a relationship.

Community stability depends on an informal social contract—an unwritten, unspoken agreement to act well and behave honestly. We seem to be especially sensitive to breaches of the social contract: humans are often poor at solving abstract logical puzzles, but we have no trouble at all correctly solving exactly the same logical task when it is presented as a social dilemma. We are also sensitive to being found out: for many people, guilt and embarrassment are strong motivators for toeing the social line—so much so, in fact, that it seems that simply placing a pair of eyes in a public space is enough to reduce both petty theft and littering (the *watching eyes* phenomenon). And we have a mechanism (gossiping) for ensuring that knowledge about backsliders' bad behavior gets around the network. These all form part of the protective mechanism that we humans have evolved to prevent deceivers and freeriders from destroying the integrity (and size) of our social groups.

Because life in stable social groups depends on trust more than anything else (I trust that you won't exploit me), we use lots of cues as fast-and-furious guides to trustworthiness, most of which are cultural in origin. These all seem to be cues that identify the community we grew up in, and hence the group of people we know well enough to trust implicitly. They include having the same dialect; growing up in the same place; and having the same educational trajectory, the same hobbies and interests, the same moral/religious/political views, the same sense of humor, and the same musical tastes. They form the *seven pillars of friendship*, and along with our folk tales and social histories (see Question 91) they make up our cultural worldview of who we are and how we came to be. The more of them we share with someone, the more we trust them, the more likely they are to make a good friend, and the more altruistic we are likely to be to each other.

89. How have primate societies evolved?

Primates are one of the oldest mammal lineages, whose origins go back well before the extinction of the dinosaurs (see Question 47), although these very early archaic primates were more squirrel-like than monkey-like. After the disappearance of the dinosaurs, these early primates diversified and then, during the Eocene era, beginning around 55 million years ago, gave rise to early true primates ("primates of modern aspect"). These early primates were all prosimians and, like the modern members of this group (the lemurs of Madagascar and the galagos of the African mainland), they were nocturnal (as we can tell from their very large eyes) (see Question 63), with small bodies; they scurried and leaped about in trees (rather like squirrels); and they were probably largely solitary.

The rise of the anthropoid primates (the monkeys and apes), around 40 million years ago, was mainly associated with the adoption of a diurnal way of life, based on well-developed color vision and a switch from insectivory to leaf eating (folivory) and fruit eating (frugivory). The earliest species were still arboreal and lived in dense forest, as do many of their descendants. However, as the climate became cooler from around 20 million years ago, some lineages moved down onto the forest floor, and from there out beyond the forest into the adjoining woodlands and grasslands. Each of these moves was associated with ever-higher predation risk, larger groups, more complex social systems, and bigger brains (see Questions 81 and 85).

The evolution of social groups for all these lineages seems to have followed a very similar trajectory. Solitary individuals first formed into small multimale/multifemale groups and then in larger ones as the demands of predation risk increased. While group size remained modest, species were likely to switch back and forth between groups with several adult males and groups with just a single breeding male, depending mainly on whether or not individual males were able to successfully defend groups of females against rivals (see Question 78). However, under pressure from infanticide risk, multimale groups sometimes broke up into reproductive pairs occupying separate territories in which each female is joined by a male who acts as her bodyguard (the *bodyguard hypothesis*).

In the primates, monogamous pair-bonding, however, appears to have been a demographic and cognitive dead end in that it never seems to be possible for species that have entered into it ever to get back out and return to any of the other social arrangements. Species that go down this route seem to be stuck there forever. This is not true in either carnivores or ungulates, where switches into and back out of monogamy seem to have been common. In primates, it seems that the psychological demands of lifelong monogamy are so different that they require changes in brain circuitry that cannot easily be undone. It may be that this is because of the kinds of bonded relationships that primates had already evolved long before any of them opted for monogamy. Carnivores and ungulates, in contrast, have much more casual relationships, and so their forms of monogamy may be less intense.

Infanticide by males is a universal problem for mammals because of the lengthy period of maternal investment (gestation plus lactation) (see Question 75). However, it is a relatively minor risk for most mammals and becomes a serious issue only for species like primates that have large brains and hence very long intervals between births. So long as the baby continues to suckle, it disengages the menstrual endocrine system, causing menstrual amenorrhea (see Question 82). Across all mammals (including humans), this mechanism is determined by the rate at which the infant suckles: the key rate for the switch between fertility and infertility (and vice versa) is exactly one bout of suckling every four hours. How long the infant suckles for doesn't seem to matter.

Because of this, infanticide has been a major factor in primate social evolution. By killing dependent infants when they take over a social group, males can cause the female's reproductive endocrine system to reset; the female will start her menstrual cycles again within a few weeks and will conceive within a few months. For a male whose tenure as a breeding male is typically just two or three years, the need to wait for a year or two until a baby stops suckling represents a significant loss in his total reproductive output (and hence his fitness). The evolutionary pressure to move the female on as quickly as possible is intense, and the intensity increases as the length of the female's normal reproductive cycle (the time between one birth and the next) increases. In those species where females form coalitions, females will often gang up against an infanticidal male, although if the males are very much larger than the females this doesn't always help. It's a classic example of intersexual conflict (see Question 77).

90. Have human societies evolved?

Over the past 10 millennia, human societies have progressively moved from the classic hunter-gatherer form of fission-fusion sociality (in which a community of 100–200 individuals is distributed in three or four small camp groups, with individual families able to move between camp groups when it suits them) to settled villages, city-states, kingdoms, and eventually the kinds of nation-states in which most of us now live. The Victorians certainly thought this was a natural evolutionary sequence. Evolution, however, is almost never progressive in the way implied by this sequence, so we probably should be suspicious of any such claims.

A better way of thinking about this is in terms of the function (or adaptive value) of social groups. In almost all birds and mammals, animals live in groups as a defense against predators. The size of the group reflects the risk of predation. Historical sociologists have argued that the push toward settlement in the Neolithic, and the progressive growth in settlement size that followed, was a consequence of raiding by neighbors. In other words, other humans increasingly became the predators, replacing conventional predators in importance as we became more successful as a species and our population density increased. This forced the scattered camp groups to come together in villages where they could defend themselves against raiders more effectively. That this continued to be a problem through the following millennia is indicated by the large number of Iron Age forts in Europe:[1] these were invariably built in defensive positions, many with ditches full of arrowheads and evidence of having been destroyed by conflagration.

As with all biological processes, any change in how something works has inevitable consequences elsewhere in the system. The growth in settlement size incurs at least three major problems. One is ecological: how to feed a large, spatially concentrated population. Foraging out from a central roost inevitably denudes the environment nearest to the center, so that, with time, individuals have to forage farther and farther away, thereby exposing themselves once more to risk of attack by raiders as well as conventional predators—the very problem that settlement was designed to avoid. The solution was agriculture, or at least an intensification of the kinds of casual agriculture that hunter-gatherers were already doing at the time.

The second cost is the stress of living in close proximity (see Question 82). Living on top of each other inevitably results in an increase in petty squabbles, and even outright violence, as people become increasingly frustrated with each other. The third is a simple coordination problem: how to keep everyone on the same page, and especially how to prevent people from leaving as a result of these stresses, thereby losing the defensive benefits of having a large group. One solution was the evolution of hierarchical social systems in which a strong chief imposes both his will and a degree of political stability on the community. With this inevitably come laws and police forces.

Associated with these have been the evolution of doctrinal religions (see Question 97).

Detailed statistical analysis of the historical development of the Austronesian and Polynesian societies of the Pacific indicates that these evolved step by step along the same trajectory in different locations, moving progressively from simple village societies with no formal structure and no chiefs to simple chiefdoms, then complex chiefdoms with rituals and hierarchies of officials, and eventually fully fledged states with the equivalent of kings. These structural changes are a simple consequence of the need to manage increasing numbers of relationships as communities get larger. This much is evidenced by the fact that when states later fail and go into decline, they typically go through the same sequence in reverse as their populations decrease in size.

Some communities, like the Hutterites of North America, deliberately sidestep this problem by splitting their communities whenever they exceed about 150 individuals. They do this precisely because they want to avoid having hierarchies and laws in order to manage their affairs and relationships. By keeping community size small, they can manage their relationships by peer pressure alone. It seems that you don't necessarily have to embark on a trajectory toward the nation-state, but to avoid doing so you have to keep your community size very small. If for any reason you need to increase community size, then the evolutionary treadmill seems to be unavoidable because we do not have the social and psychological mechanisms to mitigate the stresses of living in large communities in any other way.

10

EVOLUTION OF CULTURE

91. Does culture evolve?

Culture is defined as those aspects of our behavior and beliefs that are learned—passed on from parents to children or between two unrelated individuals, by copying and instruction. Given that evolution simply means the way traits change over time, and all cultural phenomena change over time, culture has naturally been of interest to evolutionary biologists. Two well-studied examples of cultural evolution, and the changes they undergo as they are passed down from one generation to the next over many thousands of years, are folk tales and languages.

The folk tale known as "The Tale of the Kind and the Unkind Girls" (or just "The Tale of the Two Sisters") occurs in some 700 versions all over Europe. They clearly all originate from the same ancestral version several thousand years ago—conceivably even predating the arrival of the speakers of the modern European languages in Europe some 6,000 years ago (see Question 93). Around 9% of the variation in story structure is explained by differences between geographical populations (about the same proportion as we find in genetic variation between human populations), and probably reflects something very similar to genetic drift (see Question 29). As the story gets further and further from its origins in space and time, so it collects errors and modifications in the telling, and does so at a more or less constant rate. Ethnolinguistic grouping also explains some of the variation, suggesting that particular versions of the story are carried down language lineages as part of their cultural heritage as these disperse and diversify.

Languages provide the quintessential example of cultural evolution. Give or take a few arguable cases, there are about 6,000 languages spoken in the world today, and probably many times that number that are now extinct. Although these are now all mutually incomprehensible (that's the definition of a language), on the basis of similarity in their words and grammatical structures they reduce to around 60 families of related languages, and half a dozen supergroupings called *language phyla.*

That languages can be grouped in this way was first recognized by the eighteenth-century Welsh judge and philologist William Jones, who, while working in India,[1] recognized that Sanskrit shared many words in common with Latin and Greek, modern Persian, and the languages of northern Europe. On this basis, he argued that these form a single ancestral group now known as Indo-European (one of the half-dozen major language phyla). It includes most (but not all) of the languages now spoken between Europe's Atlantic coast and Bengal on the eastern side of India, including many of those spoken in modern Iran, Afghanistan, and the northern plains of India. By reconstructing its original vocabulary from words that are common throughout these languages, it has been possible to deduce that its origins lay in a group of horse-based pastoralists living on the central Russian steppe who, about 4,000 years ago, underwent a rapid expansion and swept westward to displace all the languages in Europe and southward into India and its geographical neighbors.[2]

Cultural evolution of these kinds exhibits many parallels with more conventional genetically based evolution. These include the facts that we inherit our cultural beliefs and ways of behaving from cultural "parents" (whether or not they are related to us), that cultural traits tend to "breed true" (cultural "offspring" have the same beliefs as their cultural "parents"), that they seem to behave in a quantum fashion (we inherit an entire set of beliefs just as we inherit a complete biological trait like an ear or an eye intact rather than just part of it), and that over periods of time these beliefs change as a result of copying errors (mutations) in ways very similar to genetic drift (see Question 29). These quanta of culture were named *memes* by Richard Dawkins as a parallel to the term *gene*. Although the term has fallen out of favor in evolutionary circles (mainly on the grounds that it is rather vague), it does serve a useful purpose as a term for units of culture.

Cultural evolution differs from genetic evolution in several important ways, however. It occurs much faster (cultural "generations" can be as short as a few minutes when a series of people copy each other in rapid succession, whereas, at least in humans, genetic generations are 25–30 years long). Inheritance is not always vertical (that is, from parent to offspring): it can often be horizontal (between two unrelated peers) or diagonal (between an older teacher and a biologically unrelated younger pupil). In this respect, of course, there are some striking similarities with the way viruses convey genes from one organism, or even species, to another.

Most surprising of all, perhaps, is the fact that cultural inheritance can be much more reliable than genetic inheritance. The heritability (the proportion of the variation in a trait that is due to inheritance from a parent as opposed to environmental effects) of most genetic traits is typically around 20%, whereas the heritability of some cultural traits such as religious beliefs can be as high as 70%. If you want to maintain the biological integrity of a lineage, it might, ironically, be best to have most of the inheritance done via culture rather than by genes.

92. Why are human cultures so different?

One obvious explanation is that beliefs and behaviors are adapted to local circumstances. They are ways of solving the problems of how to survive in a given environment—which plants are poisonous if eaten, which places are safe to visit, how best to hunt particular species, which trees make the best firewood or bows. The extraordinary skills of the Micronesian and Polynesian navigators allowed them to sail all over the western Pacific using knowledge of wave patterns, star patterns, and cloud formations above isolated islands, along with charts made from sticks and shells, that were passed on from master navigator to pupil. These skills enabled the populations of isolated islands to trade with each other and to find safety in times of disaster or famine. Such information wouldn't have been of much interest to non-seafaring communities elsewhere.

Cultural inheritance allows the wisdom of generations to be passed on by learning, thereby avoiding the risks, not to say tedium, of each generation having to figure out such complicated things for themselves by trial-and-error learning (with a *very* high likelihood of eating something poisonous or getting completely lost). Because people's lives depend on it, natural selection can keep this kind of cultural knowledge stable over many thousands of years. Aboriginals living along the southern coast of Australia are said to have in their folk tales a very detailed map of the sea floor of the Tasman Sea—despite the fact that it has not been dry land since the end of the last ice age some 10,000 years ago.

Some cultural phenomena, however, may serve no obvious function at all, and we can expect these to drift randomly over time, such that two populations descended from a common ancestor many millennia earlier may end up with completely different beliefs. In this respect, cultural evolution behaves in ways similar to genetic drift (see Question 29). Folk tales offer one example (see Question 91). In other cases, cultural components may coevolve because they are locked into, or jell well with, each other or some other aspect of the population's social or cultural behavior. Having a belief in high gods, for example, might result in the coevolution of a moral system that fits well with the concept of an all-seeing god.

In other cases, a new cultural system may replace a previous one wholesale, just as one invading species can replace a previous one. This has often happened with languages: Anglo-Saxon replaced Latin and Celtic in England after the Romans left, and Hungarian replaced its predecessors in Hungary after the Huns arrived. In some cases, this may be because the previous language community is wiped out, but it can also happen when the original inhabitants adopt the language of a few invaders because it is considered socially superior (e.g., the way the Gauls of France and Spain adopted Latin after being conquered by the Romans, giving rise to the modern Romance languages). Another familiar example is the adoption of a new religion, as when Christianity replaced the Norse and Germanic pagan religions of northern Europe, or Islam replaced the religions of the Near East and Africa, almost overnight through mass conversions.

Cultural phenomena, and especially languages, can remain astonishingly unchanged over time, as though they are under stabilizing selection (see Question 12). Some words, like *one* (or *me/I*), *mother, brother, fire, hand, black,* and *you,* have hardly changed at all in form and meaning over many tens of millennia and are virtually identical in many of the world's languages. Most of these are words that refer to things that everyone has or experiences, and they are typically words that are used frequently. Words that are rarely used are much more likely to change over time and distance; they are more at the mercy of fashion, behaving like neutral genes that are subject to drift. Grammar can be even more stable across languages: some grammatical forms seem to have been stable for as long as 50,000 years. This may reflect the demands of communication and the selection that this places on our ability to interact. We can invent new words, and they can be very appealing; but if we invent a new grammar, no one understands us.

Most of our culture, however, is really designed to identify who we are as a group and allow us to recognize each other the moment we open our mouths. Dialect and the "seven pillars of friendship" (see Question 88) provide a shortcut to knowing who belongs to our community, and hence whom we can trust. For this reason, the important thing is that if we do belong to different communities, my culture should be as different as possible from yours so that you are instantly recognizable as not being a member of my community. And, as long as they don't affect survival, it doesn't really matter what particular things you and I believe in, providing they are different.

93. Why have so many different languages evolved?

Language has clearly played a central role in the evolution of modern humans, and much of what we do—from culture to education, to science, engineering, and medicine—depends on it. Without language, there would be no culture. As a result, there has been a longstanding interest in both why language evolved (the purpose it serves in our lives) and when it did so. Everyone has, perhaps naturally, tended to see language as being about the transmission of information.

The general assumption has been that language functions as a medium of instruction (e.g., for toolmaking) or cooperation (mainly for hunting). In fact, neither of these particular functions actually needs language: learning how to make tools is best done by observation and practice ("Just watch me" is about as much language as you need), and hunting is best done—and most often is done—alone or in silent groups. The real problem, however, is that an information exchange function doesn't really make any sense given that there are so many (around 6,000) different, mutually incomprehensible languages in use today—never mind all the ones that have gone extinct. Nor does it make sense when we think of how easily languages fractionate. Why make it so difficult to communicate if the purpose of language is to exchange information? Why should this process happen *so* quickly—within as few as two dozen generations?

In just a millennium and a half after the collapse of the Roman Empire around 500 AD, Latin spawned a dozen descendants, most of which are now all but mutually unintelligible (Italian, French, Romanian, Spanish, Catalan, and Portuguese, plus some minor ones including Sardinian and Occitan). Similarly, linguists recognize six languages in the English family: English, Lowland Scots, Caribbean patois, Black Urban Vernacular of the United States, Krio (Sierra Leone creole), and Tok Pisin (New Guinea pidgin)—with the English spoken on the Indian subcontinent on the verge of being elevated to being the seventh. They all descend from Anglo-Saxon spoken in the eighth century AD, and, with the exception of Scots, all have been in existence for less than 400 years (and many a lot less).

There are really two questions here: why did language (in the singular, meaning grammatically structured vocalizations) evolve, and why did different languages (in the plural) evolve?

The most plausible function for language is, in fact, facilitating social cohesion (as opposed to cooperation). The central problem that humans faced throughout their evolutionary history has been how to maintain cohesion within their increasingly large communities (see Question 87). The scale of the bonding problem is evident from the fact that no other primate has groups larger than about 50, yet humans can manage groupings three times this size. Language, once it had evolved, became one of the crucial components of that process. It allowed us to tell jokes (and hence tap into the way laughter triggers the endorphin system) (see Question 87) and regale each other with folk tales and other stories that defined the community (see Question 91).

Some indirect evidence to support this is provided by what people actually talk about in conversations both in the postindustrial societies (where 60% of conversation content is social) and among hunter-gatherers. In a study of the Ju/'hoansi Bushmen of southern Africa, the American anthropologist Polly Wiessner found that most of the conversations in the evening are social stories, whereas those in the daytime are more factual. Yes, we do use language to teach, but teaching and other functions look much more like useful evolutionary by-products, a kind of icing on the evolutionary cake (sometimes known as *windows of evolutionary opportunity*). The most useful function that language provides is the transmission of information about the state of our social networks and our status on the seven pillars of friendship (see Question 84).

But even if social information exchange is language's main function, what could possibly explain the rapid diversification of mutually incomprehensible dialects and languages? The answer is differentiation between social communities—allowing me to recognize instantly that you do, or do not, belong to my group. So, once again, the answer is social.

This suggestion makes sense of two otherwise curious observations made by social linguists. The first is that, at least during the 1970s before mass media reduced the number of local dialects, it was possible to place a native English speaker to within 35 kilometers (22 miles) of his or her birthplace. In hunter-gatherer societies, an area with a diameter of 70 kilometers (44 miles) identifies the typical size of territory for the tribe: the tribe is a linguistic grouping—the community that speaks the same language.

The second observation is the widely observed fact that working-class parents often make great efforts to ensure that their daughters speak well, whereas their sons are left to speak as they like. Although inevitably interpreted in terms of the patriarchy (despite the fact that it was the mothers who were largely responsible for enforcing this), a more plausible explanation, given what we know about human marriage patterns and the importance of community, is that parents are trying to maximize their daughters' marriage opportunities: by speaking neutrally, they increase their chances of marrying up the social scale and doing better for themselves—and hence their family lineage. But boys have no such opportunities and invariably have to marry within their social class; consequently, their interests are best served by being well embedded into their local culture where they can receive support from their peers and the wider community. Hence, it is best for them to learn their local dialect and stick with it.

94. Does any species besides humans have culture?

There have been many claims over the last half century or so that a wide range of nonhuman species rival humans in having culture. This has ranged from birds prizing the stoppers off milk bottles to get at the cream during the 1950s, Caledonian crows using tools to obtain food, whales singing different songs in different places, Japanese macaques washing sweet potatoes to remove sand before eating them, and chimpanzees using tools of many different kinds to fish for insects, crack nuts, or obtain water.

The central question here is exactly what culture actually means in animals. Two major views have emerged. Field workers (mainly zoologists) typically argue that any difference in behavior between populations, or even between groups within a population, is evidence of culture. In contrast, experimentalists (mainly psychologists) argue that this definition is too lax and would allow population differences driven by individual learning in response local ecology to be included. The key for the psychologists lies in the mode of transmission: to be absolutely sure it is culture and not the result of convergence through individual trial-and-error learning, the behavior must have been passed on from one individual to another by mindless copying.

Psychologists live in fear of the "clever Hans effect." Kluge (or Clever) Hans was a horse that bewitched his schoolmaster owner as well as audiences throughout Germany in the 1900s with his apparent ability to count by tapping out with his hoof the answer to simple arithmetic questions. Careful experiments eventually revealed that he was just very good at picking up cues from his master, who was doing the actual counting and giving out very subtle cues in his breathing patterns as to when the horse had reached the right number of taps.

This hard-nosed empiricism is, of course, admirable, since we might otherwise be deceived into thinking animals (and even humans for that matter) can do something genuinely clever when in fact they are solving a problem by some very simple seat-of-the-pants method. In the case of cultural behavior, this might include more mundane psychological processes like emulation (noticing that other animals are eating something at a particular spot, and later discovering how to access the same food by trial-and-error learning). However, its downside is that we might miss a lot of behaviors that are genuinely cultural.

Here, the slippage in definition between what we mean by *culture* in animals and *culture* in humans comes to the fore. In humans, we really think of "high culture," by which we mean storytelling, drama, literature, dance, the rituals and meaning of religion, and so on. Much of this, as anthropologists never cease to remind us, involves attributing meaning to actions or things. In their view, culture is the substance of what it is to live in society, where mutually held beliefs about the world inform what we do and how we define our relationships with each other and even with the environment. The bald truth is that no animal even comes close to humans in terms of its "cultural" behavior. Indeed, it is doubtful whether any animal even comes close to what a five-year-old human child does in this respect.

Nonetheless, animals' cultural abilities, and especially those of primates, are not insignificant and should not be dismissed as of no importance. They demonstrate stages on the way to fully formed human *Culture* with a capital "C." Like the pinhole eyes of planaria (see Question 12), they show us the intermediate steps in an evolutionary sequence. They demonstrate that the mechanism of transmission that underpins human culture has deep roots, making their exaptation to produce fully human culture evolutionarily plausible. Culture did not arise by some single genetic supermutation (or *macromutation*) or even special creation, but by a series of small steps. Once again, we need to be careful not to be misled by the apparent magnitude of the differences between us and other animal species and the fact that none of the intermediate species have survived (see Question 66).

95. Why do only humans have "high culture"?

The fact that humans seem to be on a different planet from all other animals in terms of culture, just as they are in terms of language, raises the obvious question about what makes this possible. It obviously has something to do with our cognitive abilities, but which ones? And how do these relate to the underlying neurobiology and its evolution over the course of primate evolution?

Probably the most important difference lies in the cognitive capacity known as mentalizing or mindreading (see Question 69). Mentalizing is a recursive phenomenon, with a natural limit for most normal adults at five recursions of mental states: "I *believe* [1] that you *think* [2] that I *suppose* [3] that you *wonder* [4] whether I *intend* [5] . . . [something]"—with the successive mental state verbs in italics and the successive mentalizing levels indicated in brackets. The best that most sentient animals can do is first-level mentalizing ("I *believe* [something]"), although great apes all seem to be able to achieve second-level mentalizing ("I *believe* [1] that you *think* [2] . . . [something]"), the level achieved by five-year-old human children. So the fact that most adults can manage five levels (and some can do better than that) represents an enormous cognitive step up from any other animals.

Without aspiring to third-level mentalizing as a minimum, complex language and storytelling is impossible, not least because we would be able to manage only single-clause sentences of the kind that children produce when they first learn language. "Jim loves Mary" is no doubt interesting, but it is not really the basis of a sophisticated, emotionally stirring story: we wait in vain for the "and . . ." or "but . . ." that heralds the fact that something *really* interesting is going to follow. Yes, we can communicate with second-level mentalizing (young children do this perfectly well, after all), but our stories will be of a kind that are so impoverished most people wouldn't recognize them as Culture with a capital "C."

What is important for both casual language exchanges (conversations) and storytelling is that we can see the world from our audience's point of view ("I *understand* how my audience *interprets* what I am *intending* to say . . .," with three mentalizing levels). This allows us to structure what we say, and how we say it, so as to convey better just what we mean. This is especially important when we are trying to convey emotions and mental states, because language itself is actually not very good at doing this. Experimental studies indicate that people find stories and jokes that have more mentalizing levels (in effect, more characters and their mind states) much more interesting and enjoyable than stories with fewer. In short, mentalizing seems to be the key to both our complex sociality and our culture.

Perhaps the most important findings over the last decade or so, however, have come out of neuroimaging. Many thousands of brain scan studies of mentalizing have now been done, and they identify a particular set of neurons in the brain that seem to be responsible for this ability. These form an integrated network (known as the theory of mind, or mentalizing, network) that links units in the prefrontal cortex (the bit of the brain above the eyes where both conscious thinking and the meaning of emotions are processed) with units in the temporal lobes (just by our ears) via the area known as the temporoparietal junction at the back of the temporal lobe. This same network correlates with the number of friends we have (see Question

85). It turns out that monkeys also have this circuit, and it seems to play a crucial role in their social abilities too.

More importantly, brain scan studies of adult humans reveal that individuals' mentalizing competencies (how many mind states they can handle at one time) correlate with the volume of the prefrontal cortex in particular. It is this part of the brain that has increased out of all proportion during the course of primate evolution and that is absolutely (even if not relatively) much larger in humans than in any other species. In fact, there is a three-way correlation between a person's mentalizing capacities (how many levels they can manage without getting confused), the size of their social circle, and the size of these key brain regions.

In short, it is these advanced mentalizing abilities, and the specialized neurobiological network in the brain that underpins these, that make it possible for us to have complex language, tell sophisticated stories, have high culture, and, it seems, manage many social relationships at the same time—and, of course, make sense of the fact that other animal species can't do any of these (or, if you prefer, do science as successfully as we do).

96. When and why did music evolve?

Humans seem to have a natural, and universal, propensity to engage in music. We don't know exactly when humans first sang a song or played a musical instrument (though it is likely that the first happened long before the second). There is archaeological evidence for musical instruments dating from 35,000 years ago in Germany (several playable flutes made from vulture wing bones) and a possible flute made from a cave bear femur dated to 40,000 years ago (though the claim that this is any kind of musical instrument has been questioned). Since the wing bone flutes are well-worked instruments, it is clear that the earliest ones must have long predated these. Singing, handclapping, and perhaps rhythmic banging on logs are probably much older accompaniments to communal dancing, as they still are among traditional hunter-gatherers who lack musical instruments of a more conventional kind.

While many animals, notably birds, "sing," these tend not to be communal activities and are usually performed alone. They are best explained as territorial signals (some monkeys) or mate advertisement (most birds, and perhaps the famous song of the humpback whale). Some primates, such as gibbons and South American howler monkeys, engage in duets or group roars, often because of the proximity of neighboring groups: these could be either territorial "keep away" displays or group-bonding ("let's stick together") choruses. Some birds, such as the African bell bird, have calls (hardly songs, since they involve only two notes, one given by each bird in such close succession that it sounds like one bird calling) that are probably associated with pair-bonding—or at least allowing the pair to keep track of where their mate is while they are foraging.

Humans exhibit most of these functions in their singing and dancing activities. Rock bands, and even classical musicians, unquestionably attract more than their fair share of sexual interest. And the famous *haka* of the New Zealand Maori that is now performed before international rugby matches (and as a greeting for royal visitors) is in fact a war dance designed to terrify the enemy (hence some of its more exotic facial expressions). It may be that singing as a social activity evolved out of these communal forms of vocal advertising. However, it is equally likely that it evolved out of laughter or some other form of chorusing round the campfire that formed part of a group-bonding ritual.

Human music-making differs in one important respect from that of all other animals. It is a very social activity, and we almost always do it in company as a group. It seems that human music-making mainly has a social function, and that function is almost certainly a community-bonding one. Experimental studies of singing and dancing have shown that these activities both raise pain thresholds (a proxy for endorphin activation) (see Question 87) and increase the feeling of bondedness to those with whom we are singing or dancing, even if they are strangers. Importantly, the activity does not affect the sense of being bonded to close friends who are not actually present. There is an immediacy to it.

Given all this, one likely suggestion is that music-making evolved to allow early humans to bond their increasingly large communities in contexts where more personalized forms of social bonding such as social grooming were ineffective because of the numbers of people involved (see Question 87). This suggestion was in fact originally made by the French sociologist Émile Durkheim in the early 1900s, but was largely forgotten until a century later when experimental evidence for a social-bonding function based on endorphins began to appear.

A best guess for the point of origin would have been the appearance of archaic humans around 500,000 years ago when an increase in brain size suggests that social community size underwent a correspondingly dramatic increase. That makes it likely that Neanderthals sang and danced at least as well as us. In contemporary societies, this function is perhaps best exemplified by the singing of national anthems and the kinds of grandstand singing found at sporting events like football matches.

97. Did religion evolve?

Most of the major world religions (Christianity, Islam, Buddhism, Hinduism, Sikhism, Zoroastrianism, and their various derivatives) have well-documented, recent historical origins. These are usually referred to as the doctrinal religions because they are founded on explicit theological beliefs of some kind. While these may have been preceded by earlier phases of organized religion (the religions of ancient Egypt, Greece, and Rome or those of the Mayas and the Incas), most of these seem to date only from the last 5,000 years at most. The consensus among historians of religion is that these doctrinal religions were preceded by a range of shamanic-type religions similar to those still found among hunter-gatherer societies. These hunter-gatherer religions are mainly "religions of experience": they typically lack theologies, gods, priests, religious-based moral codes, formal rituals, and places of worship, and instead focus on trance states induced by song and dance, psychotropic drugs, or other behaviors at social gatherings.

Most of these shamanic-type religions are ultimately concerned with social bonding: they are communal activities that are never undertaken alone. The singing and dancing, and the trance states that are triggered by these, stimulate the release of endorphins in the brain, which in turn give rise to an enhanced sense of community bonding. Among bushmen, trance dances take place at irregular intervals, usually because community relationships are becoming fractious. The psychosomatic release induced by trance (most likely via the endorphin system) seems, in effect, to wipe the hard drive clean and reset community relationships back to the default state.

In an important sense, all religions demarcate the community of believers as a self-contained social group who belong together and have a particular set of beliefs about a transcendental world (the seven pillars of friendship again) (see Question 88). Along with the particular rituals that a community practices for accessing the spirit world, these beliefs neatly differentiate one community from all other communities that hold different beliefs. In this way, religious practices reinforce the bonds within the community by emphasizing the fact that our beliefs are different from everyone else's beliefs. In many ways, it is the ultimate form of us-versus-them mantra. This form of religion may be quite ancient. In contrast, doctrinal religions may be much more recent.

Doctrinal religions seem to have evolved in the Neolithic: this, at least, is when we first see buildings that seem to be religious spaces ("temples") and evidence for symbolism, ritual, and hierarchy. The most likely reason these religions appeared at this point is that something more robust was needed to mitigate the costs of living in spatially restricted communities where tensions build up quickly. In hunter-gatherer societies, these stresses (and their effect on female fertility) (see Question 82) are mitigated by fission-fusion sociality: if the stresses start to become too much, individuals can move between camp groups without ever having to leave the community as a whole. The novel problem faced in the Neolithic was that the whole community of 150 or so individuals was forced to live together in the confined space of a permanent village. Nothing is a better recipe for rapidly escalating aggravations, mayhem, and murder.

Doctrinal religions seem to provide the perfect carrot-and-stick solution that would have kept the lid on frayed tempers just enough to prevent the community from breaking up at the first signs of stress—for two reasons. First, the more mystical aspects of religions and their rituals provided an emotional reason to sign up to the community project, just as they had always done in hunter-gatherer societies. Second, doctrinal religions are usually associated with a moral code and some kind of omniscient moralizing high god who could see what mere mortals could not and so punish backsliders. Strictly speaking, a moralizing high god isn't essential: a doctrine of reincarnation such as that found in some of the eastern religions does the same job through its threat that misdemeanors in this life will be punished by a return in the next life as a lower form of life.

The advantage of this kind of threat is that no one can test its validity until it is too late. Thus, a belief in some kind of moralizing force instilled during childhood could effectively reduce (albeit not eliminate) the level of antisocial behavior experienced by a community. Nineteenth-century American utopian communities provide some evidence of how effective religion can be in keeping a community together: communities survived around 10 times longer if they had a religious basis than if they were strictly secular. Without the religious dimension, people fell out with each other more quickly, and the community fell apart as a result. More importantly, this is a problem that will escalate as the size of the community increases. Discipline imposed by a moral police force is then almost always the only way of ensuring conformity.

98. Will the internet change the course of human evolution?

Evolution never stops. So any new circumstance could, in principle, influence the course of human evolution; the one thing it definitely won't do is bring evolution to a stop. As a forum for the exchange of knowledge, the internet is most likely to influence our future through cultural rather than biological evolution through its capacity to greatly speed up the spread of information. In this, it could be a force for good (by spreading useful knowledge) or evil (by spreading false rumors).

Part of the problem with the digital world is that users, and especially users with less experience of life, risk information overload. At the touch of a button, we can find out more about more in a matter of minutes than we could have managed in many years, perhaps even decades, of face-to-face interaction. The problem is that we don't have the skeptics standing at our shoulder pouring cold water on outrageous claims. It's an environment that leads to "echo chamber" effects. At the best of times, we prefer opinions that reinforce our own views. We can't ignore the skeptic at our elbow in the face-to-face world, but we can defriend them at the push of a button in the digital world.

There is one context in which this might become a problem, and that is in respect of children's development. Earlier, we saw that the number of friends you can have is a function of your brain size and your endorphin receptor density (see Question 86). But these only set the parameters of the board game and are not of themselves sufficient to produce a functional human adult capable of negotiating their way through the complexities of the human social world. Fine-tuning these skills requires a great deal of practice and experience, even in monkeys and apes.

In humans, that fine-tuning seems to take up most of the first 25 years of our lives. It is a long, slow haul in which we learn how to cope with individuals of radically different social skills and objectives; we learn how to negotiate, how to hone our inhibitory skills (see Question 83), and how to balance short-term gains against their long-term consequences. That requires a very long period of direct exposure to other people in contexts where you just *have* to learn how to compromise and how to take the rough with the smooth because you cannot escape.

The danger may be that, if a large part of this learning period is spent online rather than in the face-to-face world, these diplomatic skills may not be learned so effectively. That could mean that children will grow up less capable of coping with the vagaries of the adult social world and will have smaller rather than larger social circles as a result. That sounds like a recipe for social breakdown rather than greater integration and community cohesion. Unfortunately, it will take a generation before we find out whether or not this prognosis is true, and by that time we may have a generation of socially less skillful parents who will pass on their own limitations to their children.

99. Does the theory of evolution have implications for disciplines outside biology?

Humanities scholars and social scientists have sometimes asserted that there is a moment where evolution ends and history begins. History is usually taken to mean that our behavior is no longer subject to brute biology but is more strongly influenced by social mores and cultural beliefs inherited by learning. However, such a view drives a wedge between evolution and human behavior that doesn't actually exist.

The mistake seems to be the result of assuming that animals' behavior is completely genetically determined. But it is genetically determined in only the very simplest life forms. Brains evolved as a way of buffering the individual from the vagaries of the environment, allowing it some flexibility in how to exploit opportunities or avoid threats (see Question 60). The capacity to learn is crucial. It is the goal state at which the animal has to aim and the capacity to make decisions to achieve this that are genetically determined, not its behavior. The problem, of course, is that the culture-is-history argument confuses *inheritance mechanisms* with evolutionary *function* (see Question 10). Cultural processes are just as much under the influence of evolutionary forces as anything else (see Question 91).

Economics is another discipline where evolutionary theory might be relevant. Natural selection is mathematically an optimization process, and optimization is a central assumption in economics. The only difference between an evolutionary approach and an economic approach is that the latter typically focuses on financial payoffs whereas evolutionary biologists emphasize fitness (the genetic payoff). Money is often a direct proxy for fitness because we use it to buy resources or invest it in our offspring. The big question is whether using fitness rather than an artificial proxy like money might conceivably improve the quality of microeconomic forecasting by focusing directly on the criteria that actually motivate human behavior.

The historical sciences (history, archeology) are all about human behavior and the decisions we make in life. Understanding the reasons that motivate human behavior, and the social and cognitive constraints that act on this, might provide us with new kinds of insights into why historical events took particular turns. Does a society collapse, for example, because it exceeds its members' capacities to manage their relationships? Here's one example of how evolutionary can illuminate an historical phenomenon. Medieval Icelandic Viking *berserkers* had higher fitness than more collegial individuals, offering an explanation why such behavior persisted (irrespective of whether or not there is a genetic predisposition to behave aggressively) (see Question 78). Analyses of the historical data have shown that the Vikings adhered closely to Hamilton's rule (see Question 25). They were much less likely to murder a close relative than a distant one, and, if they did, it was always because the prize gained was much greater: unrelated individuals might be murdered over a trivial insult, but close relatives would be murdered only to gain a kingdom.

Literature is often seen as beyond the evolutionary pale, but the evolved psychology of storytellers and their audiences has much to tell us about why stories are constructed in particular ways, how and why we respond to them in the way we do, why we enjoy some stories more than others, and why the whole business of storytelling evolved in the first place. Politics and the law are all about communities trying to come to some kind of *modus operandi* for living together. These often involve tussles, sometimes violent tussles, between different ways of achieving this, as well as between the conflicting interests of different individuals or cliques. We need these formal arrangements only because self-interest keeps floating up to the surface in all human interactions. The natural evolutionary conflict between selfish genes and altruistic individuals would also seem to be directly relevant to theology with its preoccupation with good and evil and how to make sense of, and regulate, humans' sometimes destructive behavior.

100. So why do people still misunderstand the theory of evolution?

This is a very good question. One problem seems to be that Darwin's theory of evolution often challenges our ideologies, and humans never react entirely rationally when their ideologies are challenged. In many cases, the objections to evolutionary theory are based on genuine misunderstandings about what the theory claims, often because critics confuse different senses of Tinbergen's "four whys" (see Question 10).

One particularly infamous misunderstanding is known as the junkyard paradox. In essence, it asserts that evolution cannot be true because random changes simply couldn't produce species as we see them: it would be about as likely as a tornado blowing through a junkyard producing a fully assembled jumbo jet. However, a tornado in a junkyard is not an appropriate metaphor for how evolution works. Nor is it even a reasonable description of how a jumbo jet came to be. Natural selection is a cumulative process that builds progressively on what has gone before, and does so over very long periods of time. The correct analogy here, as Darwin himself recognized, is that of the human designer selecting, just as pigeon fanciers do, progeny that are better and better adapted to the form they (or the environment) have in mind. This is exactly how aircraft designers produced the jumbo jet—adding bits to existing aircraft models, trying out new designs, and changing the designs if they didn't work.

Another classic fallacy is that if evolution is gradual, there simply cannot have been enough time to produce all the finely adapted species that we find. This would indeed be true *if* it was the case that evolution accumulated new traits only piecemeal entirely at random through genetic drift (see Question 29). However, when under selection, traits can in fact change surprisingly fast (see Question 16). Granted, it takes longer to produce an entirely new species than just to change a single trait like lactose tolerance or sickle cell anemia, but even so, we are talking about only a few hundred millennia. In fact, the average lifespan of a species in the human lineage (the time between its first and last appearances) is only about half a million years. The problem with this claim is that it confuses selection with the rate of mutation.

Another claim that is that humans, at least, don't behave in order to promote their genes, but because they are motivated to do something. I marry someone because I fall in love with them, not in order to contribute genes to future generations. This is certainly true, but once again it confuses two different levels of explanation. There is an important distinction between the level at which evolution operates (the genetic consequences of behavior, namely fitness) and the mechanisms through which natural selection achieves these ends (motivations). Natural selection works through motivations because only motivations have the causal immediacy to persuade us to act: the genetic consequences of our actions are simply too remote to provide any organism with the motivation to do anything. In effect, natural selection tunes these motivations to achieve the desired fitness goals—or, to put it more correctly, natural selection results in the evolution of those motivations that maximize fitness.

Another classic claim is that the theory of evolution is circular: it assumes evolution in order to prove that evolution is true. This is rather an odd claim, because nowhere in the theory of evolution, as presented by Darwin or anyone since Darwin, is evolution assumed at the outset. What is assumed at the outset (axiom 1) (see Question 4) is that animals (or species) vary—a simple observable fact. Evolution (meaning the change that occurs in a species' appearance or form) is the outcome, or consequence, of a number of mechanistic processes represented by the other two assumptions (the axioms of adaptation and inheritance). There is nothing intrinsically evolutionary about any of these axioms. The misunderstanding seems to have arisen because fitness as a genetic property (essentially the fourth concluding statement) has been confused with the role of genes as the mechanism of inheritance (axiom 3) (see Question 4).

Though less frequently heard now, another common error has been to ask where are the "missing links"—the half-chimpanzee/half-human midway point between the two living species. The very concept of a missing link, however, is misleading. Any two species (say, chimpanzees and humans) are the end products of separate evolutionary trajectories from their common ancestor; the human lineage did not evolve from chimpanzees. Their common ancestor could, in fact, have been very different from both of them. As it happens, the rate of change in the chimpanzee's genome has actually been much faster than in ours, despite the fact that we look like we have undergone more evolutionary change.

Group selection (the claim that evolution occurs for the benefit of the species) is another old chestnut. Evolution does not occur for the benefit of the group or the species; it occurs only for the benefit of the genes. Group selection is impossible simply because it is, in effect, a form of genetic altruism and so would always be undermined by selection at the level of the individual. It has sometimes been claimed that kin selection is a form of group selection, but this isn't true. If anything, kin selection is a form of group-level selection (see Question 81), and that is a very different beast, not least because the fitness accounting takes place at the level of the gene as in all Darwinian processes.

Evolutionary theory has often been equated with Social Darwinism, the late nineteenth-century political movement that has been partly blamed for the rise of Nazi political philosophy in the early twentieth century. In fact, despite its name, Social Darwinism had nothing to do with Darwin (indeed, Darwin disapproved of it). It was the invention of the nineteenth-century political philosopher Herbert Spencer, with a little help from Darwin's cousin Francis Galton (the eminent geneticist and founding father of the eugenics movement). Neither Social Darwinism nor eugenics are direct consequences of Darwinian evolutionary theory. Indeed, given that they both aspire to genetic purity (i.e., genetic homogeneity), they are radically anti-Darwinian: under Darwinian evolutionary theory, the more genetic variation you have the better, as this is the essential motor of evolutionary success and you can never tell which variants are going to be most successful in the future. Genetic homogeneity is very bad news indeed: as conservation biology has repeatedly warned, lack of genetic variation is a recipe for rapid extinction.

One final misunderstanding has more to do with the philosophy of science. Evolution and biology are sometimes castigated for being post hoc descriptions of the world rather than theoretically driven sciences like physics. This is to misunderstand both science and evolutionary biology. The theory of evolution by natural selection provides an elegantly simple overarching theory whose *application* in the biological world leads to enormous complexity and diversity *because* that biological world is complex and multidimensional. Moreover, the point of any historical science is, as all those who study historical phenomena from cosmology to archeology and conventional history know only too well, to be able to explain the past—how the world came to be the way it is. In the philosophy of science, this is called *postdiction*. It is always best if we can predict things we hadn't expected, but providing principled explanations for something is just as good. More importantly, evolutionary theory often does predict things in the real world that we had not anticipated. Much of the experimental work in optimal foraging theory (see Question 12) does just that: it predicts how animals should behave *if* they are optimizing their fitness, and tests whether that is what they actually do. It seems that they do.

Perhaps the biggest problem with all of these misunderstandings is that they focus on particular issues. What they ignore is the fact that all of them are embedded within a complex web of interrelationships and causal explanations whose ramifications extend throughout all the intricate levels of the world we live in from the nature of species to human cultural behavior. Not only does that give the theory immense structural strength, but it makes it difficult to object to one component without being forced to find convincing alternative explanations for all the other parts too. Darwin's entangled bank is so tightly interwoven in a set of such elegantly interlocking explanations, all of which arise from his original theory, that it becomes difficult to pick its elements apart without losing intellectual coherence.

NOTES

Chapter 1

1. Chevalier is a minor French military title, roughly equivalent to knight.
2. FitzRoy was later responsible for inventing weather forecasting and the modern weather map—as another commission from the Royal Navy.
3. While a young curate (assistant to the parish priest) in a country parish, Malthus had been led to these ideas by totting up the number of births (or at least christenings) and burials in the parish, and noting that the first dramatically outstripped the second.
4. Sadly, Rosalind Franklin died tragically young and missed out when the other three were awarded the Nobel Prize for this discovery in 1962. Alfred Nobel had stipulated in his will that only living people could be awarded Nobel Prizes.
5. Just to confuse the unwary still further, there are, in fact, four different ways that biologists use the term "gene," all meaning rather different things. Biologists, however, rarely confuse them—despite the fact that they very seldom specify which meaning they intend, because this is usually perfectly obvious from the context.

Chapter 2

1. Several attempts were made during the 1960s to discredit Kettlewell (including accusing him of fraud), much to the joy of creationists who viewed the "debunking" of Kettlewell's experiments as conclusive evidence against Darwinism.

However, a lengthy, very detailed combined field and laboratory study during the 2000s by the Cambridge geneticist Michael Majerus has since completely vindicated Kettlewell's conclusions.

2. Although some eminent Christian divines in the nineteenth century did suggest that God had included them to fool gullible scientists—though it has to be said that nowhere in any of the Abrahamic scriptures is there any suggestion that God is predisposed to practical jokes. Even so, one would have thought that one such joke might have been enough. Besides, having to wait nearly 6,000 years from the date of creation in 4004 BC to spring the joke seems to suggest remarkable patience as well as prescience on the part of the Creator—it would have been a dreadfully elaborate waste of time had no one ever thought up the theory of evolution in the meantime!

Chapter 3

1. Fitness is formally defined as how much better or worse a trait (or gene) does compared to the average for the population. Conventionally, we compare the performance of some new mutant trait against the average for the normal trait (whose fitness is usually taken to be 1).

2. Technically, he termed this *neighborhood modulated fitness* because he calculated it as the fitness of the gene distributed across all members of the population. However, he showed that this gene's-eye view was virtually identical to calculating an individual's inclusive fitness using his simple equation.

3. In many ways, his insight is much more important than this. He argued that Hamilton's conception of inclusive fitness implied that if we want to maximize our fitness, we should be interested in the fate of the currently pubertal generation (the generation about to reproduce). Not only do I share my interests in them with my genetic kin, but I also share them with my in-laws because they have an interest in their success as well, and so I should take account of their interests too. A study of human social networks confirmed that we do indeed treat in-laws almost exactly as if they were genetic kin, just as Hughes's insight predicts.

Chapter 5

1. Most modern fish belong the family of ray-finned fishes. Lobe-finned fish were, until around 300 million years ago, one of the

most abundant members of the fish family, and are thought
to have been the stock from which all living land vertebrates
evolved. Their fins are attached to the body by a bony stump that
extends from the skeleton, which enables some of their living
descendants (the lungfish) to crawl across the muddy bottom
of the shallow African lakes where they now live; unusually for
fish, they are also able to survive out of water by gulping air into
a primitive lung formed out of part of their swim bladder. Like
the coelacanth, lungfishes are very large (typically two meters
[six foot] long), though none of the living representatives of this
family come close to the seven-meter (23 foot) length of some of
their extinct fossil relatives.

Chapter 6

1. Named after the seventeenth-century Cambridge ostler Thomas
   Hobson, who became famous for allowing those who rented
   his horses to choose whichever one they liked providing it was
   the one that had been in his stable longest (and was hence most
   rested)—in other words, no choice at all. A Hobson's Choice
   strategy is, thus, one where there really is no choice.

Chapter 7

1. This in itself is odd: all other species of animals have just a single
   species of body louse.
2. The hominins are the members of our lineage, including both the
   australopithecines and members of the genus *Homo*. The term
   hominids (with a "d") was used for this grouping in the older
   literature, but this has now been expanded to include all the
   great apes.
3. Well into the eleventh century AD, British slaves were being
   traded by the Vikings as far as southern Russia and Turkey.

Chapter 8

1. Despite having been a major force in American psychology in the
   closing decades of the nineteenth century and *the* major advocate
   of a role for evolution in psychology, Baldwin has been all but
   forgotten, mainly thanks to his dismissal from Johns Hopkins
   University, Maryland, in 1908 for having been discovered in
   a brothel—an event that provided his many enemies with an
   opportunity to get at him. After being sacked, he retired to
   France, where he died in 1934 having done no more science. This

single episode caused the future course of psychology to take a radically different direction in which evolution played no role. Ironically, he is now much better remembered for his contribution to biology.

Chapter 9

1. Iron Age: roughly 1100 BC to 400 AD.

Chapter 10

1. He was employed as a judge by the East India Company, responsible for adjudicating civil cases within the territories that the Company managed. Since many cases were disputes between Indians that had to be judged according to local practice, he decided that he needed to learn Sanskrit (the ancient language in which all such law was written) in order to be able to do his job properly. It was then that he realized that many words in Sanskrit were very similar to the equivalent words in ancient Greek.

2. The only pre-Indo-European language that survived in Europe was Basque, mainly because its speakers were able to escape into their mountain retreat in the Pyrenees. The other modern non-Indo-European languages (Finnish, Hungarian, and Estonian) are, by comparison, recent arrivals, having been brought in after the fall of the Roman Empire as part of a long series of invasions from eastern Europe by yet more nomads (Huns, Mongols, Tartars, and others) from Central Asia.

FURTHER READING

1. EVOLUTION AND NATURAL SELECTION

Coyne, Jerry (2010). *Why Evolution Is True*. New York: Oxford
  University Press.
Desmond, Adrian & Moore, James (1994). *Darwin*. New York: Norton.
Dunbar, Robin (1995). *The Trouble with Science*. London: Faber & Faber.
Maynard Smith, John (1993). *The Theory of Evolution*.
  Cambridge: Cambridge University Press.
Prum, Richard (2017). *The Evolution of Beauty: How Darwin's
  Forgotten Theory of Mate Choice Shapes the Animal World—and Us*.
  New York: Doubleday.
Tinbergen, Niko (1963). On aims and methods of ethology. *Zeitschrift für
  Tierpsychologie*, 20: 410–433.

2. EVOLUTION AND ADAPTATION

Lents, Nathan (2018). *Human Errors: A Panorama of Our Glitches, from
  Pointless Bones to Broken Genes*. New York: Houghton Mifflin
  Harcourt.
Lister, Adrian (2018). *Darwin's Fossils: The Collection That Shaped the
  Theory of Evolution*. Washington DC: Smithsonian Books.
Ruxton, Graeme; Allen, William; Sherratt, Tom & Speed, Michael (2004).
  *Avoiding Attack: The Evolutionary Ecology of Crypsis, Warning Signals
  and Mimicry*. Oxford: Oxford University Press.
Stevens, Martin (2016). *Cheats and Deceits: How Animals and Plants
  Exploit and Mislead*. Oxford: Oxford University Press.
Williams, George (2018). *Adaptation and Natural Selection*. Princeton,
  NJ: Princeton University Press.

Williams, George (2013). *Plan and Purpose in Nature*.
London: Weidenfeld & Nicholson.

3. EVOLUTION AND GENETICS

Carey, Nessa (2015). *Junk DNA: A Journey Through the Dark Matter of the Genome*. New York: Columbia University Press.

Dawkins, Richard (2016). *The Selfish Gene*. Oxford: Oxford University Press.

Hughes, Austin (1988). *Evolution and Human Kinship*. Oxford: Oxford University Press.

Meneely, Philip; Hoang, Rachel; Okeke, Iruka & Heston, Katherine (2017). *Genetics: Genes, Genomes, and Evolution*. Oxford: Oxford University Press.

Nettle, Daniel (2009). *Evolution and Genetics for Psychology*. Oxford: Oxford University Press.

Ridley, Matt (1999). *Genome: The Autobiography of a Species in 23 Chapters*. London: Fourth Estate.

Williams, Gareth (2019). *Unravelling the Double Helix: The Lost Heroes of DNA*. London: Weidenfeld & Nicolson.

4. EVOLUTION OF LIFE

Dyson, Freeman (2010). *Origins of Life*. Cambridge: Cambridge University Press.

Lane, Nick (2006). *Power, Sex, Suicide: Mitochondria and the Meaning of Life*. Oxford: Oxford University Press.

Levy, Elinor & Fischetti, Mark (2007). *The New Killer Diseases: How the Alarming Evolution of Germs Threatens Us All*. New York: Crown.

Nesse, Randy (2019). *Good Reasons for Bad Feelings: Insights from the Frontier of Evolutionary Psychiatry*. New York: Dutton.

Zuk, Marlene (2008). *Riddled with Life*. New York: Harvest Books.

5. EVOLUTION OF SPECIES

Benton, Michael (2015). *When Life Nearly Died: The Greatest Mass Extinction of All*. London: Thames & Hudson.

Brusatte, Steve (2018). *The Rise and Fall of the Dinosaurs: The Untold Story of a Lost World*. London: Macmillan.

Carson, Rachel (2002). *Silent Spring*. New York: Houghton Mifflin Harcourt.

Darwin, Charles (2004). *On the Origin of Species by Means of Natural Selection of the Preservation of Favoured Races in the Struggle for Life*. *[1859]*. London: Routledge.

Diamond, Jared (2013). *The Rise and Fall of the Third Chimpanzee.*
New York: Random House.

Grant, Peter & Grant, Rosemary (2011). *How and Why Species
Multiply: The Radiation of Darwin's Finches.* Princeton, NJ: Princeton
University Press.

Mayr, Ernst (1982). *The Growth of Biological Thought: Diversity, Evolution,
and Inheritance.* Cambridge, MA: Harvard University Press.

Wilson, Edward (2001). *The Diversity of Life.* Harmondsworth: Penguin.

6. EVOLUTION OF COMPLEXITY

Beukeboom, Leo & Perrin, Nicolas (2014). *The Evolution of Sex
Determination.* Oxford: Oxford University Press.

Hansell, Michael (2007). *Built by Animals: The Natural History of Animal
Architecture.* Oxford: Oxford University Press.

Montgomery, David & Biklé, Anee (2015). *The Hidden Half of Nature: The
Microbial Roots of Life and Health.* New York: Norton.

Pimm, Stuart (1991). *The Balance of Nature? Ecological Issues in the
Conservation of Species and Communities.* Chicago, IL: University of
Chicago Press.

Ridley, Matt (1994). *The Red Queen: Sex and the Evolution of Human
Nature.* Penguin.

7. EVOLUTION OF HUMANS

Dunbar, Robin (2014). *Human Evolution: Our Brains and Behavior.*
Harmondsworth: Pellican (in USA: Oxford University Press).

Gamble, Clive; Gowlett, John & Dunbar, Robin (2014). *Thinking
Big: How the Evolution of Social Life Shaped the Human Mind.*
London: Thames & Hudson.

Reich, David (2018). *Who We Are and How We Got Here: Ancient DNA and
the New Science of the Human Past.* Oxford: Oxford University Press.

Roberts, Alice (2015). *The Incredible Unlikeliness of Being: Evolution and
the Making of Us.* London: Heron Books.

Rutherford, Adam (2018). *A Brief History of Everyone Who Ever Lived: The
Human Story Retold Through Our Genes.* London: Weidenfeld &
Nicolson.

Stringer, Chris (2012). *Lone Survivors: How We Came to Be the Only
Humans on Earth.* London: Macmillan.

8. EVOLUTION OF BEHAVIOR

Alcock, John & Rubenstein, Dustin (2019). *Animal Behavior.*
New York: Sinauer.

Barash, David & Lipton, Judith (2009). *Strange Bedfellows: The Surprising Connection Between Sex, Evolution and Monogamy*. New York: Bellevue Literary Press.

Birkhead, Tim (2000). *Promiscuity: An Evolutionary History of Sperm Competition*. Cambridge, MA: Harvard University Press.

Cheney, Dorothy & Seyfarth, Robert (2008). *Baboon Metaphysics: The Evolution of a Social Mind*. Chicago, IL: University of Chicago Press.

Wrangham, Richard & Peterson, Dale (1996). *Demonic Males: Apes and the Origins of Human Violence*. New York: Houghton Mifflin Harcourt.

Wyatt, Tristram (2017). *Animal Behaviour: A Very Short Introduction*. Oxford: Oxford University Press.

9. EVOLUTION OF SOCIALITY

Dunbar, Robin & Shultz, Susanne (2017). Why are there so many explanations for primate brain evolution? *Philosophical Transactions of the Royal Society, London*, 244B: 201602244.

Christakis, Nicholas (2019). *Blueprint: The Evolutionary Origins of a Good Society*. Boston, MA: Little, Brown.

Johnson, Allen & Earle, Timothy (2000). *The Evolution of Human Societies: From Foraging Group to Agrarian State*. Stanford, CA: Stanford University Press.

Marshall, James (2015). *Social Evolution and Inclusive Fitness Theory: An Introduction*. Princeton, NJ: Princeton University Press.

Wilson, Edward (2019). *Genesis: On the Deep Origin of Societies*. London: Allen Lane.

10. EVOLUTION OF CULTURE

Cronk, Lee (1999). *That Complex Whole: Culture and the Evolution of Human Behavior*. Boulder, CO: Westview Press.

Dunbar, Robin (1996). *Grooming, Gossip and the Evolution of Language*. London: Faber & Faber.

Fitch, Tecumseh (2010). *The Evolution of Language*. Cambridge: Cambridge University Press.

Henrich, Joe (2015). *The Secret of Our Success: How Culture Is Driving Human Evolution, Domesticating Our Species, and Making Us Smarter*. Princeton, NJ: Princeton University Press.

Laland, Kevin (2018). *Darwin's Unfinished Symphony: How Culture Made the Human Mind*. Princeton, NJ: Princeton University Press.

Mesoudi, Alex (2011). *Cultural Evolution: How Darwinian Theory Can Explain Human Culture and Synthesize the Social Sciences*. Chicago, IL: University of Chicago Press.

Mithen, Steven (2006). *The Singing Neanderthals: The Origins of Music, Language, Mind and Body*. London: Weidenfeld & Nicolson.

Rachels, James (1991). *Created from Animals: The Moral Implications of Darwinism*. Oxford: Oxford University Press.

Richerson, Peter & Boyd, Robert (2008). *Not by Genes Alone: How Culture Transformed Human Evolution*. Chicago, IL: University of Chicago Press.

Ridley, Matt (1997). *The Origins of Virtue*. Harmondsworth: Penguin.

Ruhlen, Merritt (1994). *The Origin of Language: Tracing the Evolution of the Mother Tongue*. New York: Wiley.

INDEX

For the benefit of digital users, indexed terms that span two pages (e.g., 52–52), may, on occasion, appear on only one of those pages.